PORTLAND PRESS RESEARCH MONOGRAPH

Structural and Dynamic Properties of Lipids and Membranes

PORTLAND PRESS RESEARCH MONOGRAPH III

Structural and Dynamic Properties of Lipids and Membranes

Edited by
Peter J. Quinn
Richard J. Cherry

Portland Press
London and Chapel Hill

Published by Portland Press Ltd, 59 Portland Place, London
W1N 3AJ, U.K.
In North America orders should be sent to Portland Press Inc.,
P.O. Box 2191, Chapel Hill, NC 27515-2191, U.S.A.

© 1992 Portland Press Ltd

ISBN 1 855 78 014 3   ISSN 0964-5845

British Library Cataloguing in Publication Data
A catalogue record for this book is available from the British Library

All rights reserved
Apart from any fair dealing for the purposes of research or private study, or criticism or review, as permitted under the Copyright, Designs and Patents Act, 1988, this publication may be reproduced, stored or transmitted, in any forms or by any means, only with the prior permission in writing of the publishers, or in the case of reprographic reproduction in accordance with the terms of licences issued by the Copyright Licensing Agency. Inquiries concerning reproduction outside those terms should be sent to the publishers at the above-mentioned address.

Although, at the time of going to press, the information contained in this publication is believed to be correct, neither the authors nor the editors nor the publisher assume any responsibility for any errors or omissions herein contained. Opinions expressed in this book are those of the authors and are not necessarily held by the editors or the publishers.

Typeset by Unicus Graphics Ltd, Horsham, Sussex and
Printed in Great Britain by Whitstable Litho Printers Ltd

Contents

	Preface	vii
	Abbreviations	ix
1	Has a biophysical approach provided insight into lipid and membrane structure? A personal perspective D. Chapman	1
2	Protein-mediated cholesterol absorption by small intestinal brush border membranes G. Lipka, D. Imfeld, G. Schulthess, H. Thurnhofer & H. Hauser	7
3	Unsaturation and the interactions of phospholipids with cholesterol and proteins K. M. W. Keough	19
4	Modulation of membrane lipid phase behaviour by chemical modification *in situ* P. J. Quinn	29
5	Cubic and hexagonal lipid phases G. Lindblom & L. Rilfors	51
6	The interaction of membrane-intrinsic proteins with phospholipids J. C. Gómez-Fernández & J. Villalaín	77
7	Cholesterol–phospholipid interactions and the exchangeability of cholesterol between membranes M. C. Phillips	103
8	Developments in the nuclear magnetic resonance spectroscopy of lipids and membranes: the first quarter century E. Oldfield	119

9	Rotational diffusion of membrane proteins: studies of band 3 in the human erythrocyte membrane using triplet probes R. J. Cherry	137
10	Modulation of protein structure by the lipid environment W. K. Surewicz, A. Muga & H. H. Mantsch	153
11	Infrared spectroscopic studies of membrane proteins D. C. Lee	165
12	Biocompatible surfaces based upon biomembrane mimicry Y. P. Yianni	187
	Subject index	217

Preface

In a rollcall of scientists who have made significant contributions to our knowledge of biological membranes, the name of Dennis Chapman is amongst the foremost. This book aims to honour that achievement and is authored entirely by his former students and colleagues. One of the remarkable features of the book is that we have been able to produce a volume that is balanced broadly across the topics of membrane structure and dynamics and this reflects the fact that his interests and influence have been spread widely. This is quite an achievement in a subject that was for many years divided into lipidologists on the one hand, who tended to regard anything that was insoluble in chloroform/methanol as devoid of interest, and proteinologists on the other, whose primary preoccupation was the removal of contaminating lipids.

Dennis Chapman, December 1991

Dennis Chapman's pioneering approach was to employ the biophysical tools of the chemist to tackle what at the time was considered to be the monumentally complex structure and dynamics of biological membranes. His interest in polar lipids dominated the early days of his career and it was then that ideas of fluidity, phase transitions etc. were developed as a result of penetrating studies using infrared spectroscopy, differential scanning calorimetry, magnetic spin resonance techniques, flash photolysis and electron microscopy. It is rather ironic that, having first employed vibrational spectroscopy to probe the structure of lipids, he has exploited the same technique more recently to such effect in the examination of membrane protein structure.

His early stint in industry demonstrated that the creation of a laboratory able to achieve fame through its academic achievements was also able to contribute to commercial enterprise. He has been one of the few scientists to have made a successful transition from industry to academe or, as is perhaps more correct to say, felt entirely at ease in both cultures. His achievements have been marked by the award of many distinctions, not least of which was his election to the Fellowship of the Royal Society. This volume, which is based on a colloquium held at Royal Holloway and Bedford New College in December 1991, is also a tribute to Dennis Chapman's contribution to the subject of biological membranes and their constituents.

Peter J. Quinn
Richard J. Cherry
London 1992

Abbreviations

ACTH	Adrenocorticotropic hormone
BBM	Brush border membrane
BBMV	Brush border membrane vesicle
CP	Cross-polarization
DAPC	Diacetylenic phosphatidylcholine
$DC_{8,9}PC$	1,2-Bis(tricosa-10,12-diynoyl)-sn-glycerol-3-phosphocholine
DGalDG	Digalactosyldiacylglycerol
DGlcDG	Diglucosyldiacylglycerol
DMPC	Dimyristoylphosphatidylcholine
DMPG	Dimyristoylphosphatidylglycerol
DNPC	1,2-Dinonanoyl-3-phosphocholine
DOPE	Dioleoyl-phosphatidylethanolamine
DPCC	Dipalmitoylphosphatidylcholine
DPG	Diphosphatidylglycerol
DPH	1,6-Diphenyl-1,3,5-hexatriene
DPPC	1,2-Dipalmitoyl-sn-phosphocholine
DSC	Differential scanning calorimetry
EPC	Egg phosphatidylcholine
FRAP	Fluorescence recovery after photobleaching
GAPDH	Glyceraldehyde-3-phosphate dehydrogenase
GAPlaE	Glycerol acetal of PlaE
H_I	Hexagonal phase
H_{II}	Reversed hexagonal phase
I	Isotropic phase
LaLPC	Lauroyllysophosphatidylcholine
L_α	Lamellar liquid-crystal phase
MAS	Magic angle sample spinning
MGalDG	Monogalactosyldiacylglycerol
MGlcDG	Monoglucosyldiacylglycerol
MO	Mono-olein
MRI	Magnetic resonance imaging
PA	Phosphatidic acid
PC	Phosphatidylcholine
$Pd(QS)_2$	Palladium di(sodium) alizarine monosulphonate
PE	Phosphatidylethanolamine
PG	Phosphatidylglycerol

PI	Phosphatidylinositol	
PlaE	Plasmaenylethanolamine	
POPC	1-Palmitoyl,2-oleoyl-*sn*-phosphocholine	
PS	Phosphatidylserine	
PVC	Polyvinylchloride	
SEM	Scanning electron microscopy	
SM	Sphingomyelin	
SUV	Small unilamellar vesicle	
TEG	Thromboelastography	
T_{LH}	L_α to H_{II} phase transition	
ToFSIMS	Time-of-flight secondary ion mass spectroscopy	
XPS	X-ray photoelectron spectroscopy	

Has a biophysical approach provided insight into lipid and membrane structure? A personal perspective

D. Chapman

Department of Protein and Molecular Biology,
Royal Free Hospital School of Medicine,
Rowland Hill Street, London NW3 2PF, U.K.

When I first went to Cambridge in 1960 as a Comyns Berkeley Fellow of Gonville & Caius College, Cambridge, I was keen to learn about molecular orbital theory and I had many discussions with the theoretical chemists. On one occasion one of the young scientists asked me about my own research interests. I said that I was interested in lipids. 'Lipids?', he said; 'How many atoms are there in lipids?' I replied 'Quite a few'. 'More than four?' he asked. 'Well, yes' I replied. 'More than eight?' he challenged with some surprise in his voice. 'Well,' I said somewhat apologetically, 'yes, I'm afraid so'. 'Oh dear,' he replied; 'Dennis, you are wasting your time working on those molecules'!

For a short while I thought that he might be right. I therefore started work on a molecule, S_4N_4, and a relative S_2N_2. Indeed, Alan Massey and myself [1], using e.s.r. spectroscopy showed, I believe for the first time, that electron delocalization could occur [3] with an inorganic ring molecule S_4N_4. This led to other studies on $[SN]_x$ polymer, charge alternation and molecular orbital calculations [2]. I began to realize, however, that studies of lipid molecules and particularly phospholipid molecules would enable me to become involved in the field of biomembranes and molecular biology and so I decided to continue and to extend my interest in lipid research.

At this time in the early 1960s, in England, very few scientists were interested in lipids and phospholipids. Biochemists, in general, threw the material down the sink whilst some physicists scorned these molecules, exclaiming 'Aren't they just fats?'!

The few exceptions among biochemists interested in phospholipids at this time were Dr. Rex Dawson and Dr. Alec Bangham. Only a few pure phospholipids had been synthesized by organic chemists such as Malkin at

Bristol. The organic chemists determined the capillary melting points of the phospholipids and showed that they were independent of the chain length and unsaturation of the fatty acids making up the lipid molecules.

I decided that we should begin a programme of research in which we would synthesize pure phospholipids and characterize their physical behaviour using a range of physical techniques including calorimetry and spectroscopic methods such as n.m.r. and i.r. spectroscopy. I hoped that this could give a firm foundation for providing insight into the role of these lipid molecules in biomembrane systems. I first initiated a single crystal X-ray study of a pure phospholipid, dilauroyl phosphatidyl ethanolamine; this was later completed by G. Shipley [3]. Earlier studies with long chain anhydrous soap systems such as sodium stearate and palmitate, using i.r. spectroscopy [4], had shown that considerable melting of the lipid chains could occur sometimes hundreds of degrees below their capillary melting point (~ 300 °C). I reasoned that phospholipid molecules would also show similar phase behaviour with a transition dependent upon chain length and unsaturation. We introduced calorimetric and n.m.r. techniques to study this and to reveal the extent of molecular motion in the systems above this critical phase-transition temperature. We then related this thermotropic phase behaviour of a phospholipid to its behaviour in solids, lipid–water systems and monolayer systems [5–7].

At this time the wealth of information being revealed by chromatographic methods concerning the chain length and unsaturation of the lipid chains associated with biomembranes was considerable and somewhat overwhelming in detail. Each biomembrane showed a range of chain length and unsaturation. This distribution differed from one biomembrane to another. There were speculations at the time concerning the specific interactions between particular chains or unsaturated chains and proteins but there was too much detailed information for the picture to be seen clearly. Our studies of phospholipid dynamics led us to formulate the concept of fluidity of biomembranes.

We formulated the concept of fluidity [6, 8] as follows: 'We envisage that one of the functions of the distribution of fatty acid residues observed with these phospholipids is to provide the correct fluidity at a particular environmental temperature so as to match the required diffusion or rate of metabolic processes for the tissue. Thus in membranes where metabolic and diffusion processes are required to be of a rapid nature, such as in the mitochondria, the average transition temperature of the phospholipids present will probably be low compared with the biological environmental temperature, while in membranes where these processes are slow, e.g. in myelin of the central nervous system, the average transition temperature for the phospholipids will be higher and may be close to that of the biological environmental temperature.' We also pointed out that the lipids of poikilothermic organisms varied with the temperature of growth, suggesting that 'poikilothermic organisms and bacteria appear to have a feedback mechanism linked to the environmental temperature which enables the fatty acid residue of the phospholipids to be altered so as to keep the hydrocarbon chain fluidity fairly constant'.

With this in mind we next extended our studies to the effect of cholesterol on the dynamic motion of the lipid chains. We showed that the well known condensation effect of cholesterol observed in monolayer systems was not limited to lipids containing 9:10 *cis* double bonds but also occurred in those containing *trans* double bonds and even in fully saturated lipids [9]. Calorimetric studies showed that cholesterol at high molar concentrations would remove the lipid phase transition [10]. We indicated that there was no particular 'complex' of lipid and cholesterol. Recent discussions are consistent with this view.

Our n.m.r. studies of the molecular motion of the lipid chains in the solid state above the phase transition temperature showed particularly narrow line widths [11, 11a]. This led us to sonicate the lipid and to show proton high resolution n.m.r. spectra even with a simple 60 MHz Varian n.m.r. spectrometer [12, 12a]. There was much discussion as to why such good resolution was obtained. We followed this work by additional studies using the technique of spinning the lipid at the magic angle (so that no sonication was required) and we were again able to achieve good high resolution spectra [13]. Together with E. Oldfield, we then introduced studies of deuterium n.m.r. spectroscopy for the study of lipid dynamics [14].

At Sheffield University at the instigation of Dr. Gonzalez-Rodriguez we next turned our attention to the study of protein rotational diffusion. R. Cone had shown with rhodopsin ($r \sim 20\ \mu s$) that such protein rotational diffusion occurred. We decided to examine the membrane protein bacteriorhodopsin. In this case protein rotational diffusion was shown to be extremely limited [15, 15a]. This work led to our development of triplet probes later exploited so well by R. Cherry.

We also examined the question of the protein–lipid interaction and the extent of lipid perturbation, and our introduction of the technique of deuterium n.m.r. spectroscopy was invaluable (used by J. Seelig) for rationalizing some of the confusion that existed in this area [16].

In more recent times in England considerable pressure has been placed upon Universities and their staff to focus their research programmes for practical applications. We began to think about practical applications of lipids and membranes. This led (with Peter Quinn) to the hydrogenation of biomembranes using a water-soluble homogeneous catalyst [17]. In turn this stimulated us, at the Royal Free, to produce (synthesize) phospholipid polymers, and even to polymerize liposomes and cells and to develop new haemocompatible materials based on the properties of the phosphorylcholine head group [18, 18a].

Our latest studies at the Royal Free concentrate on biomembrane protein structure, again using spectroscopic techniques, but particularly Fourier transform i.r. spectroscopy [19–21], a technique also well exploited by

H. Mantsch (Ottawa). In our laboratory we have developed qualitative and quantitative methods (with D. Lee) for determining the secondary structure of membrane proteins and we are now studying K^+ ion channel proteins using this technique.

I believe that the biophysical approach has been most valuable for understanding lipid and biomembrane structures. There is little doubt that the application of physical techniques has transformed our concept of biomembrane structure. The progression from the fixed structures seen in the early electron microscope pictures to the now accepted dynamic structures that many membranes possess is dramatic. We know that the lipid matrix can be fluid and that the lipids and proteins can undergo rotational and translational diffusion. We can understand the modulating effect of cholesterol and the way in which proteins interact with the surrounding lipid. Now we are beginning to obtain more and more information about the secondary structure of proteins, leading to a detailed knowledge of how molecules and ions are transported through the membrane.

In all of these studies of phospholipids and biomembranes I have been fortunate to have excellent young scientists working with me, particularly at the Unilever Frythe Laboratory and later at Sheffield and London Universities. Some of these scientists have generously contributed to the chapters in this Volume. There are many others, however, who have also made important contributions to the subject. I wish to thank them all for their support and their partnership in our many explorations of these lipid biomembrane phenomena. We have had fun exploring these areas together. Now let us see what tomorrow brings.

Finally, one collaborator who was essential to me for all the scientific work and for providing a happy and stable family background is my wife, Margaret Chapman. I wish to pay full tribute to her for her constant support and encouragement.

References

1. Chapman, D. & Massey, A. G. (1962) Trans. Farad. Soc. 58, 1291–1298
2. Chapman, D. & McLachlan, A. D. (1963) Trans. Farad. Soc. 59, 2671–2679
3. Hitchcock, P. B., Mason, R., Rhomas, K. M. & Shipley, G. G. (1974) Proc. Natl. Acad. Sci. U.S.A. 71, 3036
4. Chapman, D. (1958) J. Chem. Soc. 152, 784–789
5. Chapman, D. & Collin, D. T. (1965) Nature (London) 206, 189
6. Chapman, D., Williams, R. M. & Ladbrooke, B. D. (1967) Chem. Phys. Lipids 1, 445–475
7. Chapman, D. & Salsbury, N. J. (1966) Trans. Farad. Soc. 62, 2607–2621
8. Chapman, D., Byrne, P. & Shipley, G. G. (1966) Proc. R. Soc. A290, 115–142
9. Chapman, D., Walker, D. A. & Owens, N. F. (1966) Biochim. Biophys. Acta 120, 148–155
10. Ladbrooke, B. D., Williams, R. M. & Chapman, D. (1968) Biochim. Biophys. Acta 150, 333–340
11. Salsbury, N. J. & Chapman, D. (1968) Biochim. Biophys. Acta 163, 314–324
11a. Veksli, Z., Salsbury, N. J. & Chapman, D. (1969) Biochim. Biophys. Acta 183, 434–446
12. Chapman, D. & Penkett, S. A. (1966) Nature (London) 211, 1304–1305
12a. Chapman, D., Fluck, D. J., Penkett, S. A. & Shipley, G. G. (1968) Biochim. Biophys. Acta 163, 255–261
13. Chapman, D., Oldfield, E., Doskocilova, D. & Schneider, B. (1972) FEBS Lett. 25(2), 261–264
14. Oldfield, E., Chapman, D. & Derbyshire, W. (1971) FEBS Lett. 10(2), 102–104

15. Razi-Naqvi, K., Gonzalez-Rodriguez, J., Cherry, R. J. & Chapman, D. (1973) Nature (London) 245, 249-251
15a. Behr, J. P., Chapman, D. & Razi-Naqvi, K. (1974) Biochem. Soc. Trans. 2, 960-962
16. Chapman, D., Gomez-Fernandez, J. C. & Goni, F. M. (1979) FEBS Lett. 98, 211-223
17. Chapman, D. & Quinn, P. J. (1976) Proc. Natl. Acad. Sci. U.S.A. 73, 3971-3975
18. Johnston, D. S., Sanghera, S., Pons, M. & Chapman, D. (1980) Biochim. Biophys. Acta 602, 57-69
18a. Bird, R. le R., Hall, B., Hobbs, K. E. F. & Chapman, D. (1989) J. Biomed. Eng. 11, 231-234
19. Haris, P. I., Fidelio, G. D., Austen, B. M., Lucy, J. A. & Chapman, D. (1987) Biochem. Soc. Trans. 15, 1129-1131
20. Vilalain, J., Gomez-Fernandez, J. C., Jackson, M. & Chapman, D. (1989) Biochim. Biophys. Acta 978, 305-312
21. Lee, D. C., Haris, P. I., Chapman, D. & Mitchell, R. C. (1990) Biochemistry 29(39), 9185-9193

Protein-mediated cholesterol absorption by small intestinal brush border membranes

G. Lipka, D. Imfeld, G. Schulthess*, H. Thurnhofer† and H. Hauser‡

Laboratorium für Biochemie, Eidgenössische Technische Hochschule, ETH Zentrum, 8092 Zürich and *Spital Limmattal, 8952 Schlieren, Switzerland

Summary

We present evidence here that cholesterol absorption by brush border membrane vesicles (BBMVs) prepared from rabbit small intestines is protein-mediated [1–3]. It is a second-order reaction, the mechanism of which involves collision-induced transfer of cholesterol. Cholesterol absorption is most efficient from bile salt micelles; using this kind of donor, cholesterol absorption is characterized by half times of the order of seconds. After proteolytic treatment of BBMVs the rate of cholesterol absorption is greatly reduced, becoming a first-order reaction involving the passive diffusion of cholesterol through the aqueous medium. This treatment releases into the supernatant a water-soluble lipid-exchange protein with a molecular weight of 13 kDa, which catalyses the exchange of cholesterol and phosphatidylcholine between two populations of small unilamellar vesicles (SUVs). BBMVs prepared from human small intestines behave similarly to those made from rabbit small intestines. The 13 kDa protein liberated from rabbit small intestinal brush border membrane can be purified to homogeneity by a two-step procedure involving gel filtration on Sephadex G-75 SF and either affinity chromatography or cation-exchange chromatography. The purified 13 kDa protein from rabbit small intestine has been used to raise polyclonal antibodies in guinea pig. At sufficiently high concentrations of the IgG fraction, the antibody inhibits the lipid exchange between two populations of SUVs catalysed by the 13 kDa rabbit or human protein. The anti-rabbit antibody cross-reacts immunologically with the human 13 kDa protein indicating a rather high degree of similarity between the rabbit and the

† Present address: University of California, San Francisco, School of Medicine, Cardiovascular Research Institute, San Francisco, CA 94143-0130, U.S.A.
‡ To whom correspondence should be addressed.

human 13 kDa protein. The IgG fraction of the guinea pig antiserum exhibits a partial inhibitory effect on cholesterol absorption by BBMVs when taurocholate-mixed micelles are used as donor particles. The fast phase of cholesterol absorption characterized by half times of the order of seconds is unaffected. In contrast, the slow phase of cholesterol absorption operative after about 60 s is inhibited by the antibody.

Cholesterol uptake by brush border membrane vesicles

The results of studies on the absorption of cholesterol by BBMVs are summarized in Table 1. BBMVs were prepared routinely, according to the method of Hauser et al. [4] from frozen rabbit small intestines, as well as from human small intestines. The production of mixed micelles and SUVs as donor particles and the proteolytic treatment of BBMVs with the aim of producing supernate-proteins were carried out as described previously [3]. Cholesterol absorption by BBMVs from various donor particles and lipid exchange between two populations of SUVs were performed as detailed in previous publications [1–3]. The amount of radioactive cholesterol in either egg phosphatidylcholine (EPC), SUVs or mixed micelles of lysoEPC/EPC/ cholesterol (60:38:2, by wt.) as donor particles decreased exponentially when these donors were incubated with BBMVs. From a linearization of these exponential decays pseudo-first-order rate constants were derived (Table 1). For both kinds of donor particles the pseudo-first-order rate constants k_1 increased linearly with the lipid weight ratio of acceptor:donor as expected for a second-order reaction. Furthermore, by using double-logarithmic plots of initial reaction rates as a function of either acceptor or donor concentration it was shown that cholesterol absorption is a first-order reaction in the brush border membrane (acceptor) concentration as well as the donor concentration [1, 3]. It can be concluded that cholesterol absorption by BBMVs is a second-order reaction overall.

After proteolytic treatment of BBMVs the measured pseudo-first-order rate constants for cholesterol absorption were significantly reduced (Table 1). The values thus obtained were comparable or smaller than the first-order rate constants measured for cholesterol exchange between two populations of SUVs [5]. The kinetics of cholesterol absorption from taurocholate mixed micelles are shown in Fig. 1. This figure and a comparison of the results summarized in Table 1 emphasize the special role cholate mixed micelles play as donor particles in the absorption of cholesterol. Taurocholate mixed micelles behaved quite differently from other donor particles including lysoEPC/EPC mixed micelles. As is evident from Fig. 1 and Table 1, the rate of cholesterol absorption from taurocholate mixed micelles was much faster than from any other donor particle. Under comparable experimental conditions (lipid weight ratio of BBMV to donor particle of 3.3, by wt.), the absorption of cholesterol from both EPC, SUV and lyso-

Table I. Uptake at room temperature of radiolabelled cholesterol by brush border membrane vesicles from different donor particles*

Donor	Acceptor	k_1 (h^{-1})	$t_{1/2}$ (h)	Reaction order
Unilamellar vesicles of EPC	BBMV	2.65	0.26	2
Unilamellar vesicles of EPC	BBMV; papain treated	0.080	8.7	1
LysoEPC/EPC mixed micelles**	BBMV	5.3	0.13	2
Taurocholate-mixed micelles	BBMV	1 × 10^4	6.9 × 10^{-5} = 0.25s	2
Taurocholate-mixed micelles	BBMV; proteinase K treated	0.63	1.1	1

*The ratio of brush border membrane lipid to donor lipid was 3.3 (w/w) for all measurements listed.
**Mixed micelles consisting of lysoEPC/EPC/cholesterol in a ratio of 60:38:2 by wt.

EPC/EPC/cholesterol mixed micelles (60:38:2, by wt.) was characterized by half times of the order of 10 min, while the cholesterol absorption from taurocholate mixed micelles was at least 10^3 times faster (Fig. 1, Table 1). Further, whereas the amount of radiolabelled cholesterol in EPC, SUV and

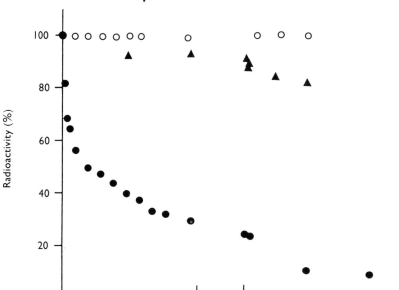

Fig. 1.   Time course of cholesterol absorption by BBMVs at room temperature

Sodium taurocholate mixed micelles (taurocholate/oleic acid/monooleoylglycerol/cholesterol at 88:6:3.6:2.6, by wt.) containing a trace of [^{3}H]cholesterol were used as donor particles. Donor and acceptor were both suspended in buffer A (0.01 M Hepes adjusted with Tris to pH 7.3, 0.3 M D-mannitol, 5 mM EDTA, 0.02% NaN$_3$) and mixed so that the final lipid concentration of taurocholate micelles was 3 mg total lipid/ml and that of BBMVs 10 mg total lipid/ml. The amount of radiolabelled cholesterol remaining in the mixed micelles was determined by scintillation counting. ●, cholesterol uptake by untreated BBMVs; ▲, cholesterol uptake by BBMVs after proteinase K treatment. The open symbols give the uptake of [^{14}C]cholate from taurocholate mixed micelles under the same experimental conditions.

lysoEPC/EPC/cholesterol mixed micelles as donor particles decreased exponentially, the decrease in radiolabelled cholesterol in taurocholate mixed micelles was not a simple exponential function (Fig. 1). The initial very fast decay in radioactivity observed in the first 10 s (Fig. 1) can be fitted by the sum of two exponentials. The two k_1 values thus derived were $k_1 = 10^4$ h^{-1}

and $k'_1 = 2.8 \times 10^2$ corresponding to half times $t_{1/2}$ of 0.25 s and 8 s, respectively.

In a double-labelling experiment taurocholate mixed micelles containing trace quantities of both [^{3}H]cholesterol and sodium [^{14}C]cholate were used. As shown in Fig. 1 the absorption by BBMVs of sodium cholate was negligible compared with that of cholesterol under these conditions [3]. Also included in Fig. 1 is the absorption of cholesterol by BBMVs that had previously undergone proteolytic treatment. After proteinase K treatment of BBMVs, cholesterol absorption from taurocholate mixed micelles was greatly reduced and was characterized by half times of the order of 1 h [3] (Table 1). Furthermore, after proteinase K treatment cholesterol absorption was mechanistically different from the protein-mediated process; it was a true first-order reaction and as such independent of BBM (acceptor) concentration.

Purification and characterization of lipid-exchange proteins

Proteolysis of BBMVs not only abolished the efficient cholesterol absorption, but also liberated a significant amount of membrane proteins into the supernatant [1–3], which are referred to as supernate-proteins. Papain treatment or, alternatively, incubation of BBMVs at temperatures above 0 °C have similar effects as to the release of membrane proteins [1–3]. Using spin-labelling these proteins were shown to bind cholestane and phosphatidylcholine (PC) [1, 2]. Furthermore, they catalysed the exchange of both cholesterol and PC between populations of SUV [1–3]. The active proteins present in the supernatant were subjected to the treatment summarized in the scheme of Fig. 2. The purification shown in this scheme consists essentially of two steps; (1) gel filtration on Sephadex G-75 and (2) affinity chromatography on Nucleosil-PC. For the latter column material, silica particles were derivatized with propylamine to which dimyristoyl PC was covalently linked [3, 6]. Gel filtration of supernate proteins on Sephadex G-75 yielded reproducibly three peaks if the eluate was analysed for PC-exchange activity (data not shown). The Sephadex G-75 column was calibrated with marker proteins so that elution volumes could be converted to apparent molecular weights. The values thus obtained for the apparent molecular mass of the proteins eluted in the three peaks were >70 kDa, 22 ± 2 kDa and 13 ± 1.3 kDa in the order of increasing elution volumes. The first peak was eluted in the column void volume and hence only a lower limit of the molecular weight could be derived. Proteins present in peak 1 were pooled, concentrated and applied to a calibrated Sephacryl S-200 HR column. An active lipid-exchange protein was eluted as a single peak corresponding to an apparent molecular weight of 100 ± 10 kDa [7]. SDS-PAGE of peaks 1–3 eluted from the Sephadex G-75 column revealed that proteins of peak 3 were least contaminated. Half a dozen proteins were recovered from this peak and these were pooled, concentrated and subjected to h.p.l.c. on Nucleosil-PC. The main portion of the protein (90–95%) was eluted as a pass-through peak (Fig. 3) that showed no lipid-exchange activity. The protein retarded on the Nucleosil-PC column

Fig. 2.   **Scheme depicting the purification of supernate-proteins**

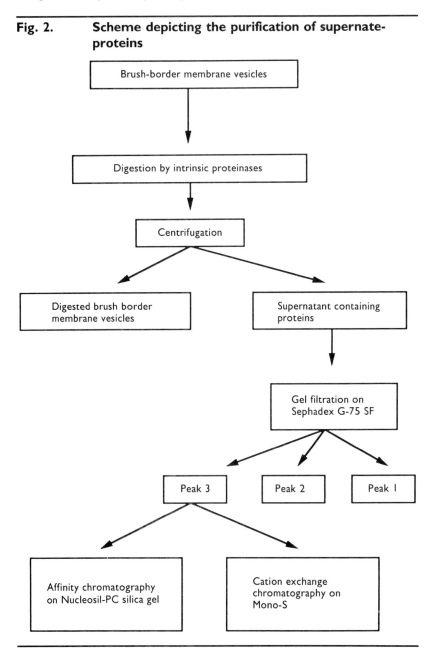

was eluted with a CHAPS gradient as shown in Fig. 3. This protein gave a single band on SDS-PAGE and was active as a cholesterol and PC transfer protein. The specific exchange activity of the purified protein was increased by a factor of about 80. The apparent molecular weight of this protein was

Fig. 3. Purification of proteins present in peak 3 by affinity chromatography of supernate proteins on a Nucleosil-PC h.p.l.c. column

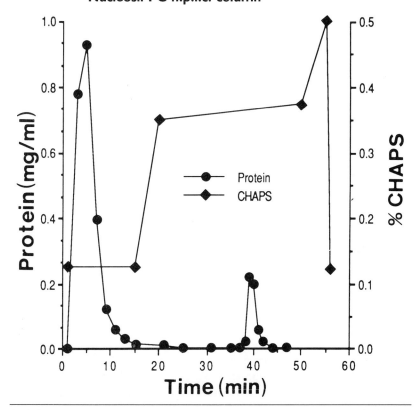

Peak 3 proteins were produced from supernate-proteins by gel filtration on Sephadex G-75. The h.p.l.c. column (10 × 0.46 cm) was run with 0.01 M sodium phosphate buffer pH 7.3, 0.14 M NaCl, 2.5 mM EDTA, 0.02% NaN_3 and 0.125% CHAPS at a pressure of 60–70 bar. Protein loading, 3.5 mg in 2 ml buffer; flow rate, 0.7 ml/min.

determined by gel filtration on Sephadex G-75, Bio-Gel P10 and SDS-PAGE. The values thus obtained ranged between 11.5 and 14 kDa with an average molecular weight of 12.7 ± 1 kDa (Table 2). Isoelectric focusing of the purified protein yielded two bands with isoelectric points of 9.1 and 9.4. The basic character of the protein prompted an alternative approach to the purification of peak 3 (see Fig. 2). Proteins present in this peak were purified

Table 2. Apparent molecular weight of the supernate protein liberated from BBMVs and purified to homogeneity, as described in the scheme of Fig. 2

Method	Apparent molecular weight (kDa)
Gel filtration on Sephadex G-75 SF	11.5 ± 1.3
Gel filtration on Bio-gel P10	13.7
SDS-PAGE	13.0

by cation-exchange chromatography using Mono-S f.p.l.c. The active protein retained on the column (0.01 M Hepes buffer pH 7.3 containing 0.3 M mannitol, 5 mM EDTA and 0.02% NaN$_3$) was eluted with a NaCl gradient at about 0.18 M NaCl. The protein thus obtained was pure by SDS-PAGE using both Coomassie blue and silver staining (data not shown).

BBMVs prepared from human small intestines behaved in a similar manner to those from rabbit small intestines. Proteolysis by either extrinsic proteinases or by activation of intrinsic ones abolished the active cholesterol uptake and at the same time a significant proportion of the total membrane protein appeared in the supernatant. The supernate proteins thus obtained also behaved similarly to those derived from rabbit small intestines. They catalysed cholesterol and PC exchange between two populations of SUV. Supernate proteins obtained from human small intestines were purified in the same way as described in Fig. 2.

Effect of antibodies against the 13 kDa protein on lipid exchange

The 13 kDa protein from rabbit small intestines which was purified to homogeneity was used to raise polyclonal antibodies in guinea pig. Antisera were routinely purified by cation-exchange chromatography on CM Affi-Gel Blue, and the IgG fraction thus obtained was concentrated by ammonium sulfate precipitation.

The effect of the purified antibody on PC-exchange activity of the 13 kDa protein is shown in Fig. 4. The lipid-exchange activity was progressively inhibited with increasing concentrations of IgG, and at IgG concentrations in excess of 6 mg protein/ml, the PC exchange was totally blocked. The PC exchange between two populations of SUV catalysed by human peak 3 proteins was also inhibited by the anti-rabbit antibody, indicating that the anti-rabbit antibody cross-reacts with the corresponding human antigen. Quantitatively, the reaction of the antibody with the human antigen was similar to that observed with the rabbit 13 kDa protein (Fig. 4).

The effect of the anti-rabbit IgG fraction on cholesterol absorption by BBMVs from taurocholate mixed micelles as donor particles is shown in

Fig. 5. As mentioned above, in the absence of IgG, cholesterol absorption from taurocholate mixed micelles was characterized by half times of the order of seconds. In the presence of IgG the fast phase of cholesterol absorption was not affected at all. There was, however, a clear-cut effect of the

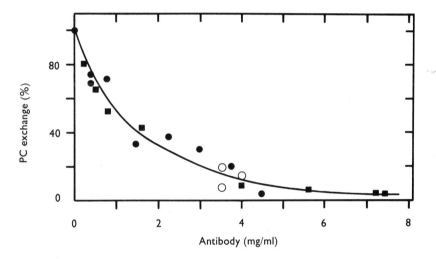

Fig. 4. **The effect of IgG on the PC exchange between two populations of SUV catalysed by the purified 13 kDa protein or by proteins of peak 3**

PC exchange was measured between two populations of SUV, the donor vesicles consisting of EPC and egg phosphatidic acid (85:15, w/w) and a trace of [^{3}H]1,2-dipalmitoyl-sn-phosphatidylcholine, and the acceptor vesicles of pure EPC. Both donor and acceptor vesicles were dispersed in buffer A and their final concentrations were 0.1 mg lipid/ml and 1 mg lipid/ml, respectively. As lipid-exchange protein, total protein of peak 3 was used. Peak 3 proteins were prepared from both rabbit and human small intestines and the effect of the anti-rabbit antibody on the PC-exchange interaction catalysed by the rabbit and human peak 3 proteins is represented by closed and open symbols, respectively. The rabbit and human antigens were used at concentrations of 32 μg protein/ml = 2.5 nM.

antibody: after 1 s cholesterol absorption was significantly slowed down and after 1 min it was inhibited completely. The residual slight decrease in radioactivity in the taurocholate mixed micelles is accounted for by passive diffusion of cholesterol from the donor particles to BBMVs. The extent of cholesterol uptake by this mechanism is similar to the cholesterol absorption of BBMVs after proteinase K treatment.

Fig. 5. Time course of cholesterol absorption by BBMVs from sodium taurocholate mixed micelles (taurocholate/oleic acid/monooleoylglycerol/cholesterol = 88:6:3.6:2.6, by wt) containing a trace of [³H]cholesterol

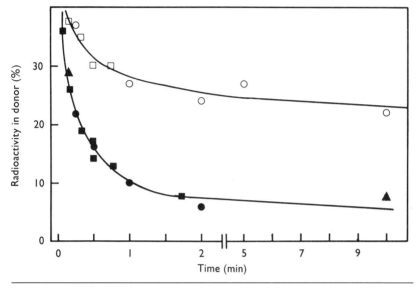

The residual radioactivity remaining in the donor was determined as a function of time at room temperature. Different symbols represent different experiments. Both donor and acceptor were suspended in buffer A (see legend to Fig. 1). The final total lipid concentrations were 3 mg/ml and 6 mg/ml for donor and acceptor, respectively. Closed symbols, cholesterol absorption in the absence of IgG; open symbols, cholesterol absorption in the presence of 3 mg IgG/ml.

Conclusions

The absorption of cholesterol by BBMVs is a second-order reaction. The kinetics of this process are consistent with a mechanism involving collision-induced transfer of cholesterol from donor to acceptor membrane. Although cholesterol absorption is a second-order reaction regardless of the nature of the donor particle, there are still significant differences depending on whether SUVs or micelles are used as donor particles. With SUVs as the donor, there is lipid exchange: one cholesterol molecule present in the donor membrane is exchanged for one lipid molecule present in the external monolayer of the bilayer of BBM, i.e., the lipid exchange is characterized by a 1:1 stoichiometry [8]. At equilibrium, cholesterol is evenly distributed between donor and acceptor lipid pools, as is expected for true mass exchange. In contrast, with micelles as the donor, there is a net transfer of cholesterol: at equilibrium practically all cholesterol is present in the lipid bilayer of BBM [9].

One of the main conclusions of this work is that cholesterol absorption by BBMVs is protein-mediated. This is true for BBMVs prepared from rabbit as well as from human small intestines. This conclusion is based primarily on the observation that the rate of cholesterol uptake is reduced drastically after proteolysis of BBMVs. Proteolysis not only affects the rate constants, but also the reaction mechanism of cholesterol absorption. After proteolysis, cholesterol absorption by BBM is a first-order reaction. The mechanism involves cholesterol desorption from the donor, diffusion of monomeric cholesterol through the aqueous phase and incorporation of cholesterol into the bilayer of BBM. Because the process is a true first-order reaction, the rate-limiting step must be the desorption of cholesterol from the donor particle. Clearly, after subjecting BBMVs to proteinase treatment, cholesterol absorption is a passive process, mechanistically quite different from the protein-mediated cholesterol uptake.

A second conclusion which can be drawn from the data presented is that cholesterol absorption by BBMVs is most efficient from taurocholate mixed micelles. This finding stresses the special role of bile salt micelles in the absorption process. As mentioned above, proteolytic treatment of BBMVs abolishes their efficient cholesterol absorption. Simultaneously, membrane proteins are liberated which behave like water-soluble lipid-exchange proteins, catalysing the exchange of both cholesterol and PC between two populations of SUV. It is reasonable to assume that the proteins released into the supernatant are part of the integral membrane protein(s) responsible for the efficient uptake of cholesterol by BBMVs. On the basis of this assumption one of the water-soluble lipid-exchange proteins was purified to homogeneity. Polyclonal antibodies against this protein were produced in guinea pig and the effect of this antibody on various activities was investigated. First, at sufficiently high concentrations of the antibody, the PC exchange between two populations of SUV catalysed by the 13 kDa antigen is completely inhibited (Fig. 4). Secondly, the PC-exchange catalysed by the corresponding human protein, i.e. the protein prepared from human small intestines, is also blocked by the anti-rabbit antibody (Fig. 4). The anti-rabbit antibody cross-reacts immunologically with the human protein. This result indicated that a protein similar to the 13 kDa protein of rabbit small intestines occurs in human small intestinal BBM. The results presented in Fig. 4 suggest a rather high degree of similarity between the two proteins.

The antibody has a partial inhibitory effect on cholesterol absorption by BBMVs when taurocholate mixed micelles are used as donor particles (Fig. 5). Apparently the fast phase of cholesterol absorption characterized by half times of the order of seconds is not affected in the presence of the antibody. The slow phase of cholesterol uptake operative after about 60 s is inhibited in the presence of the antibody. The finding that the antibody has no effect on the fast phase of cholesterol absorption is interpreted to mean that one or several different proteins are involved in this process. The relationship, if any, of the 13 kDa protein to the protein(s) responsible for the fast process of cholesterol uptake is unclear and further experiments are required to shed light on this question.

References

1. Thurnhofer, H. & Hauser, H. (1990) Biochemistry 29, 2142–2148
2. Thurnhofer, H. & Hauser, H. (1990) Biochim. Biophys. Acta 1024, 249–262
3. Thurnhofer, H., Schnabel, J., Betz, M., Lipka, G., Pidgeon, C. & Hauser, H. (1991) Biochim. Biophys. Acta 1064, 275–286
4. Hauser, H., Howell, K., Dawson, R. M. C. & Bowyer, D. E. (1980) Biochim. Biophys. Acta 602, 567–577
5. McLean, L. R. & Phillips, M. C. (1981) Biochemistry 20, 2893–2900
6. Pidgeon, C. & Venkataram, V. V. (1989) Anal. Biochem. 176, 36–47
7. Thurnhofer, H., Lipka, G. & Hauser, H. (1991) Eur. J. Biochem. 201, 273–282
8. Lipka, G., Op den Kamp, J. A. F. & Hauser, H. (1991) Biochemistry 30, 11828–11836
9. Mütsch, B., Gains, N. & Hauser, H. (1983) Biochemistry 22, 6326–6333

Unsaturation and the interactions of phospholipids with cholesterol and proteins

K. M. W. Keough

Department of Biochemistry and Discipline of Pediatrics, Memorial University of Newfoundland, St. John's, Newfoundland, Canada, A1B 3X9

Introduction

A body of information exists which indicates that the physical state of a biological membrane has a significant influence on the enzymic and transport activities performed by its constituent proteins. The most dramatic effects are seen when there are changes in lipid phase from the ordered, gel phase to the disordered, liquid-crystalline phase. Modulation of biochemical activities has also been associated with lateral phase separation in the membranes, at least by inference, if not unequivocally through demonstrated phase separations. Transbilayer asymmetry may also affect enzymic activities since it is conceivable that different phases could exist in the lipids of each half of a bilayer that surround a transmembrane protein. There is evidence that some biological activities are modulated by the viscosity or fluidity of the lipid matrix when lipids are in a single phase.

The foundations for the physical properties of the membrane lie in the structures of the individual components and how these modulate the phase and their mutual interactions. Modifications of various parts of the lipid structure, including headgroup, chain linkage, chain position, chain length and chain unsaturation, have all been shown to influence lipid phase behaviour and, either directly or by implication, to modulate biological activity (for summaries, see [1-6] for example). This chapter will present information on the influence of acyl chain unsaturation on some aspects of membrane thermotropic behaviour and selected physiological activities. It will deal with studies in model systems which can allow for simpler interpretation of the influence of lipid molecular structure on organization and activity than can be deduced from studies in whole membranes.

It has been known for some time that the introduction of a double bond into the acyl chains of a phospholipid causes a considerable reduction in the temperature (T_c) of the transition from the gel to liquid crystalline phase of bilayers formed from the phospholipid in comparison with that of bilayers formed from the saturated phospholipid with equal chain lengths [7, 8]. The decrease in the chain melting temperature caused by the introduction of a double bond means that most saturated phospholipids that are found in biological membranes would be in their liquid-crystalline state at biologically relevant temperatures. Most saturated lipids would be in the gel state at equivalent temperatures. For biologically important temperatures, an unsaturated bilayer will have a lower viscosity or greater fluidity than the equivalent saturated one. More precisely, membranes containing only saturated lipids at their respective biological temperatures would have substantially greater order and reduced molecular motion than unsaturated membranes at that temperature. Early observations on the nature of the lipids of membranes and their respective physical properties led to the concept that biological activities in natural membranes were likely to be optimal in 'fluid' lipid environments. This has been confirmed substantially by many studies of various systems (for reviews, see [3, 6, 9] for example). What has also become apparent is that it is likely that the provision of an appropriately 'fluid' environment for efficient biological activity could probably be achieved by having a relatively small number of molecular species of phospholipids in a membrane [9–11]. There are, however, hundreds, if not thousands, of individual molecular species of phospholipids to be found in biological membranes. They vary in all aspects of their structures including the degree of unsaturation of their chains (for summary see e.g. [12]).

Because the introduction of double bonds into acyl chains is an energy-requiring process it seems unlikely that cells would have evolved in a way that expends large amounts of energy to make and maintain the presence of large amounts of highly unsaturated lipids if they were unnecessary for cellular activity. Yet, cells do contain large amounts of polyunsaturated lipid, and so we assume that they are essential to optimal activity. From another view, if unsaturated lipid composition in membranes is manipulated, substantial changes in biological activities can be induced. For example, it has been found that replacement of the endogenous phosphatidylcholine (PC) of the outer monolayer of erythrocytes with PC of greater (or lesser) unsaturation leads to modified cell permeability, osmotic fragility and cholesterol efflux [13, 14].

Thus, we assume that special compositions of lipids found in individual membranes are not just accidents of evolution, but have evolved because of some specific benefit to the membrane in which they are found (e.g. [9, 11]).

Double bonds and physical properties of phospholipids

As noted above it is well known that the introduction of a double bond into one or both chains of a phospholipid causes a dramatic reduction in its

Table 1. Effect of introduction of double bonds into phosphatidylcholine on transition temperature

Phosphatidylcholine	Transition temperature (°C)	Transition enthalpy change (kcal/mol)	Reference
18:0–18:0	54.9	10.6	[44]
18:0–18:1	6.3	5.4	[45]
18:0–18:2	−16.3	3.3	[46]
18:0–18:3, n−3	−13.0	6.6	[46]
18:0–20:4, n−6	−12.6	5.3	[46]
20:0–20:0	68.4	15	[47]
20:0–20:1	22.2	8	[47]
20:0–20:2	4.2	5	[47]
20:0–20:3, n−3	4.4	7	[47]
20:0–20:4, n−6	−6.8	3	[47]
16:0–18:0	48.5	—*	P. L. J. Matthews et al., unpublished work
16:0–18:1	−2.6	—*	P. L. J. Matthews et al., unpublished work
16:0–18:2	−19.7	—*	P. L. J. Matthews et al., unpublished work
16:0–20:4, n−6	−20.6	—*	[18]
16:0–22:6, n−3	−3(−12)**	—*	[15]
18:0–18:0	54.5	9.5	[29]
18:1–18:1	−17.6	8.0	[24]
18:2–18:2	−52.4, −57	2, 1.4	[24, 33]
18:3–18:3, n−3	−73.1	1	[24, 33]
20:4–20:4, n−6	−76.9, −69.2	1, 0.8	[24, 33]
22:6–22:6, n−3	−68.4	0.5	[33]

*Not determined.
**Hysteresis seen in heating (−3°C) and cooling (−12°C) experiments.

transition temperature (see Table 1). At temperatures above the transition temperature, chain order decreases and motion increases; thus, the lower melting temperatures of the unsaturated lipids lead to the idea that biological membranes, which are usually at temperatures above the chain melting temperatures of their lipids, contain a 'fluid' lipid environment.

The early findings led to 'counting' of double bonds and association of increasing unsaturation with increasing fluidity or decreasing order and motion. Table 1 shows that the transition temperatures do not change in a fashion which would support this approach to determining the influence of unsaturation on physical properties. It is too simplistic. As can be seen in the Table, the first double bond introduced into an acyl chain causes a marked reduction in the transition temperature of the lipid. The second bond added per chain causes much less of an effect, and subsequent double bond additions cause little further change in the transition temperatures. Indeed, depending upon the total number and the positions of the bonds in the chain, the presence of multiple double bonds can lead to membranes with transition temperatures that are higher than systems with two or three double bonds. Compare, for example, the transition temperature of 16:0-18:2 PC and 16:0-22:6 PC.

The general behaviour of the transition temperatures is consistent with results obtained by deuterium magnetic resonance spectroscopy and Raman spectroscopy [15-17]. Using differential scanning calorimetry we [18] reproducibly obtained a wide transition, consistent with that of Litman et al. [16], from dispersions of 16:0-22:6 PC, although near the temperature seen for cooling rather than heating scans using ^{2}H-n.m.r. and Raman spectroscopy [15-17]. We have not yet been able to determine a reason for this difference. The difference, however, does not detract from the general message that, in numbers beyond two per chain, double bonds cause little additional reduction in the transition temperature of a lipid. Multiple double bonds may even result in a higher transition temperature than found for a lipid of equivalent chain length but a lower degree of unsaturation.

The small effect on the transition temperature of increasing the number of double bonds beyond two per chain is consistent with earlier findings of Stubbs et al. [19]. They observed that increasing the number of double bonds per chain of PC had little influence on the indicators of order and motion obtained from diphenylhexatriene fluorescence probe measurements in the liquid crystal state. Straume and Litman [20], using a measure of fractional volume available for probe reorientation, f_v, observed that PC with highly unsaturated chains (20:4 and 22:6) had a somewhat greater value of f_v (less order) than species with less unsaturated chains. These highly unsaturated PCs, containing either a 20:4 or a 22:6 chain, also have much less cooperative gel to liquid-crystalline phase transitions [16, 18]. Thus, although 20:4 or 22:6 chains do not substantially depress the transition temperature, or increase the overall rate of chain motion in a membrane, their presence could lead to increased ability to accommodate volume changes in other components in the membrane [21].

The values in Table 1 also draw attention to the fact that the position of the double bonds along the chain can influence the transition temperature of the resulting lipid. This effect was first realized when one double bond was inserted into different positions along a chain by Barton and Gunstone [22]. They observed that a double bond near the centre of an acyl chain in a phospholipid lowered the transition temperature to the maximum extent, and that progressively smaller depressions of the transition temperature were obtained as the double bond was moved towards either the head group or the methyl terminal of the chain.

The data in Table 1 are consistent with there being some effect of position of the double bonds in polyunsaturated lipids on the transition temperature. Recently, a direct comparison of dispersions of 16:0-18:3, $n-3$ PC and 16:0-18:3, $n-6$ PC has shown that the latter has the lower transition temperature, while the former exhibits more hysteresis in its transition temperature when approached from low and high temperature [23]. This confirms directly the importance of double bond position as well as number in the thermotropic behaviour of lipids with polyunsaturated chains.

One other important point is presented by the data in Table 1. The limiting effect on the transition temperature of increasing the number of double bonds per chain in lipid containing a saturated and an unsaturated chain is also seen in PCs with two polyunsaturated chains. It is worthwhile noting, however, that dipolyunsaturated PCs have very broad transitions, in the order of 40 degrees in width [24]. It has been thought that these transition widths might reflect some special melting behaviour [24], but perhaps they are not so unusual given the fairly broad transitions observed more recently from 16:0-20:4 PC and 16:0-22:6 PC [16-18].

Phospholipid-cholesterol interactions

Cholesterol is a major component of animal cell membranes and has been the subject of intense study for many years (e.g. [25]). It is in especially high concentrations in the surface membranes, and in smaller amounts in membranes in the interior of the cell. The pattern of cholesterol concentration among cellular membranes appears to be influenced primarily by the subcellular location of the membrane, rather than by the tissue of origin (e.g., [25-27]). Early studies (for summaries see [8, 28] for example) suggested that the presence of cholesterol could modify the nature of the thermotropic phase transition experienced by membranes, so as to 'buffer' the fluidity of a biological membrane.

Some recent work suggests that there are differences in the degree to which individual lipids may be influenced by the presence of cholesterol. For example, the amount of cholesterol required to remove the calorimetrically detectable phase transition of bilayers of PC varies with the molecular species of PC involved. Although the detectable phase transition of 18:0-18:0 PC was removed by the presence of 50 molar% cholesterol in mixture with the PC, it took only 40% to remove the transition from 18:0-18:1 PC [29]. Remarkably, the detectable transitions of 18:0-18:2 PC and 16:0-18:2 PC were abolished by <17 molar% cholesterol [30]. A recent extension of these studies [18] indicates that the amount of cholesterol required to remove the transition of 16:0-20:4 PC is a little higher than that required for removal of the transition of 18:0-18:2 PC, and that substantially more cholesterol (~35 molar%) is required to abolish the transition of 16:0-22:6 PC. These effects of cholesterol on various unsaturated PCs are correlated with its condensing effects on the lipids in monolayers, and with its influence on the permeability of liposomes to glucose, erythritol and glycerol, effects observed some time ago by Demel et al. [31] and Ghosh et al. [32].

Davis and Keough [29] observed that about 40% of the enthalpy change associated with the gel to liquid-crystalline phase transition of 18:1–18:1 PC remained detectable even when equimolar amounts of cholesterol were present in PC bilayers. This observation suggested the possibility that PC with unsaturation in both acyl chains may have a qualitatively, as well as quantitatively, different interaction with cholesterol than PC which contained only one or no unsaturated chain. In a recent study it was observed that cholesterol, at up to equimolar concentrations, has little or no effect on the calorimetric phase transition of dipolyunsaturated PC [33]. These results, demonstrating relatively little effect of cholesterol on the transitions of PC with two polyunsaturated chains, were consistent with the observation that cholesterol produced little or no condensing effect in monolayers of 18:2–18:2 PC and 18:3–18:3 PC [31, 32] and no effect on the permeability of vesicles of these lipids [31]. Van Blitterswijk et al. [34] found that liposomes of 18:2–18:2 PC at 25 °C were much less susceptible to the ordering effect of cholesterol, as determined by diphenylhexatriene fluorescence polarization, than were many unsaturated lipids that also contained a saturated chain in the *sn*-1 position. Litman et al. [35] have observed that, in PC containing a saturated chain at the *sn*-1 position and unsaturated chain at the *sn*-2 position, the two chains show some independence of behaviour. This could extend to interactions with cholesterol. The idea that acyl chain position and (un)saturation may have some influence on the PC–cholesterol interaction is supported by the observations of Davis and Keough [29] that the gel to liquid-crystalline transitions of the positional isomers 18:0–18:1 PC and 18:1–18:0 PC were affected to different extents by the same amount of cholesterol. The possibility should be considered that the presence of a saturated chain in a PC may be necessary to maximize its interaction with cholesterol. Overall, the data suggest that cholesterol and PC with two polyenoic chains do not mix well, and the possibility that such systems contain domains enriched in PC or in cholesterol needs to be given due weight when considering the influence of cholesterol and physical properties in membranes.

Lipid–protein interactions

There is evidence from a large number of studies that the lipid phase can strongly effect the function of most membrane enzymes, these being relatively inactive when surrounded by gel-phase lipid and active in liquid-crystalline lipid. However, once a membrane protein is in a liquid-crystalline lipid environment, as is found in almost all membranes under physiological conditions, how well do its catalytic properties correlate with lipid composition and lipid physical properties? There may not necessarily be a correlation between activity and bulk properties of the lipid such as fluidity or microviscosity (e.g. [10]). For example, East *et al.* [36] found no correlation between the order parameter measured with a spin-labelled fatty acid probe and the activity of sarcoplasmic reticulum Ca-ATPase reconstituted in a number of different PC molecular species. Yet there is ample evidence for different lipid environments being associated with proteins in different membranes, and for differences, especially in the unsaturation of fatty acyl chains, in the membranes of the same tissue of origin in species that are adapted or acclimated (partially or fully) to different temperatures.

Lee [9] supports the view that changes in activity of the sarcoplasmic reticulum Ca-ATPase that occur because of modifications of lipid structures when the lipids are in the liquid-crystalline state follow from changes in the conformational state of the ATPase or from changing the energy difference between two conformations of the enzyme. Mitchell *et al.* [21, 37] have observed that the equilibrium and rate constants for the meta I–meta II transition of rhodopsin are different in the presence of a number of different PCs, all of which are in the liquid-crystalline state. Lipids with more polyunsaturated chains produced higher equilibrium constants because they had a greater packing free volume to allow for the expanded meta II conformation. These results suggest that changes in activity may not follow changes in overall dynamic properties such as fluidity, but rely upon other interactions between membrane lipids and proteins.

The results of Mitchell *et al.* [21, 37], which suggest a special role for highly unsaturated acids in accommodating the activity of rhodopsin, can also be considered in light of the observation of Deese *et al.* [15] that rhodopsin interacts with 16:0–16:1 PC and 16:0–22:6 PC so as to differently alter the low-frequency reorientational motion of the two lipids. Thus, the interaction of lipid and protein is 'sensed' by both components, and either 'sensor' indicates some selectivity in acyl chain unsaturation in the interaction. In a related vein, the work of Jaworsky and Mendelsohn [38] and Anderle and Mendelsohn [39] suggests that, when investigating model systems, there is some preferential interaction between Ca-ATPase and lipids of varying (un)saturation. The phase properties of lipid environments containing either 16:0–18:1 PC or 18:0–18:1 PC were more profoundly influenced by ATPase than their saturated or diunsaturated counterparts. The larger effect of the protein was on mixed-acid lipids, those that are found in the natural sarcoplasmic reticulum, and suggests that it may be necessary to use appropriate lipids of correct biological structure in order to fully understand the influence of phospholipids on lipid–protein interactions.

We have begun a twofold approach to the question of how specific molecular species may influence catalytic activity. In the first of these strategies we have reconstituted the Ca-ATPase from rabbit sarcoplasmic reticulum in single species of PC, and determined the resultant activities and activity–temperature profiles. A list of activities is given in Table 2. The

Table 2. Activities of Ca-ATPase in different lipid environments at 37 °C. (P. L. J. Matthews, E. Bartlett, V. S. Ananthanarayanan & K. M. W. Keough, unpublished work)

Reconstituting lipid	T_c (°C)	Activity (μmol/mg/min)
16:0–18:0 PC	48.5	Not determined*
16:0–18:1 PC	−2.6	5.95
16:0–18:2 PC	−19.7	1.94
16:0–20:4 PC	−20.6	0.27
16:0–22:6 PC	−3(−12)**	0.12
Native purified sarcoplasmic reticulum		4.24
Purified ATPase		9.43

*Found in other work to be very low.
**Hysteresis on heating and cooling.

activity measurements and the temperature–activity profiles suggested that the enzyme found its most 'hospitable' environment (it behaved most like the enzyme in sarcoplasmic reticulum) in 16:0–18:1 PC and 16:0–18:2 PC, and not in PCs which were either more or less unsaturated. The two lipids that produced properties most like those in the natural sarcoplasmic reticulum are the most common ones found in that membrane [40–43]. These results suggest that the enzyme has a 'preference' for lipid that most resembles the endogenous lipid, in order to achieve functionality equivalent to that found in the natural membrane.

The second approach to this question has been the investigation of the same enzyme, sarcoplasmic reticulum ATPase, from two different species, rabbit and winter flounder, where the lipid environments have similar polar head group distribution, but the polyunsaturated acyl chains are different from one another [42, 43]. The dependence of K_m on temperature was different for each of these enzymes, as seen in Fig. 1, as were the activity profiles. The activation energies and the denaturation temperatures were also different. Discontinuities in the temperature-dependence of the K_m or the denaturation temperatures could not be correlated with abrupt changes in order and motion of the bulk-phase lipid, as measured by Fourier transform infrared spectroscopy [43]. There was a correlation between the loss of activity and a denaturation event which can be seen by high sensitivity differential scanning calorimetry [42]. There may be a correlation between the behaviour of the K_m and changes in protein structure seen by circular dichroism at temperatures below the denaturation temperature that are observed by calorimetry or by enzymic assay [42, 43]. Whether or not the

Fig. 1. Dependence on temperature of K_m of Ca-ATPase of sarcoplasmic reticulum from rabbit (○) and winter flounder (□)

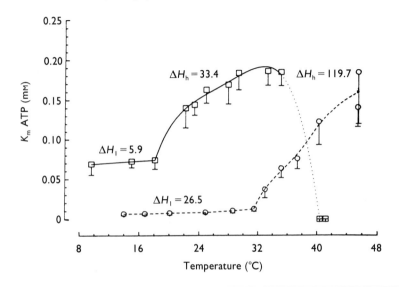

Reproduced from Vrbjar et al. [42] with permission.

changes in protein structures are in any way dependent on the lipid environment is not yet established, and 'cross-over' reconstitution experiments are underway to help to elucidate this matter.

Our own and other work suggests that there is a correlation between lipid structure and associated biological activity, and that specific lipid–protein or lipid–cholesterol interactions may be necessary for optimal function in specific membranes. Considerable additional work on carefully defined and biologically relevant model systems needs to be carried out in order to gain further appreciation of the unusual and special distribution of lipids in most membranes.

Work in the author's laboratory was supported by the Medical Research Council of Canada.

References

1. Keough, K. M. W. & Davis, P. J. (1984) in Membrane Fluidity (Kates, M. & Manson, L. A., eds.), pp. 55–97, Plenum Publishing Corporation, New York
2. Stubbs, C. D. & Smith, A. J. (1984) Biochim. Biophys. Acta 779, 89–137
3. Lewis, R. N. A. H. & McElhaney, R. N. (1992) in The Structure of Biological Membranes (Yeagle, P., ed.), pp. 73–156, CRC Press, Boca Raton, FL
4. Gruner, S. M. (1992) in The Structure of Biological Membranes (Yeagle, P., ed.), pp. 211–250, CRC Press, Boca Raton, FL

5. Selinsky, B. S. (1992) in The Structure of Biological Membranes (Yeagle, P., ed.), pp. 603-665, CRC Press, Boca Raton, FL
6. Yeagle, P. L. (1992) in The Structure of Biological Membranes (Yeagle, P., ed.), pp. 157-174, CRC Press, Boca Raton, FL
7. Phillips, M. C., Williams, R. M. & Chapman, D. (1969) Chem. Phys. Lipids 3, 234-244
8. Ladbrooke, B. D. & Chapman, D. (1969) Chem. Phys. Lipids 3, 304-367
9. Lee, A. G. (1991) Prog. Lipid Res. 30, 323-348
10. McElhaney, R. N. (1984) in Membrane Fluidity (Kates, M. & Manson, L. A., eds.), pp. 249-278, Plenum Publishing Corporation, New York
11. Keough, K. M. W. (1990) Biochem. Soc. Trans. 18, 835-838
12. White, D. A. (1973) in Form and Function of Phospholipids, 2nd edn. (Ansell, G. B., Hawthorne, J. N. & Dawson, R. M. C., eds.), pp. 441-482, Elsevier, Amsterdam
13. Kuypers, F. A., Roelofsen, B., op den Kamp, J. A. F. & van Deenen, L. L. M. (1984) Biochim. Biophys. Acta 769, 337-347
14. Child, P., op den Kamp, J. A. F., Roelofsen, B. & van Deenen (1985) Biochim. Biophys. Acta 814, 237-246
15. Deese, A. J., Dratz, E. A., Dahlquist, F. W. & Paddy, M. R. (1981) Biochemistry 20, 6420-6427
16. Litman, B. J., Lewis, E. N. & Levin, I. W. (1991) Biochemistry 30, 313-319
17. Barry, J. A., Trouard, T. P., Salmon, A. & Brown, M. F. (1991) Biochemistry 30, 8386-8394
18. Hernandez-Borrell, J. & Keough, K. M. W. (1991) FASEB J. 5, A836
19. Stubbs, C. D., Kouyama, T., Kinosita, K. J. & Ikegami, A. (1981) Biochemistry 20, 4257-4262
20. Straume, M. & Litman, B. J. (1987) Biochemistry 26, 5113-5120
21. Mitchell, D. C., Straume, M. & Litman, B. J. (1992) Biochemistry 31, 662-670
22. Barton, P. & Gunstone, F. D. (1975) J. Biol. Chem. 250, 4470-4476
23. Wassall, S. R., McCabe, M. A., Griffith, G. L., Ehringer, W. D. & Stillwell, W. (1992) Biophys. J. 61, A240
24. Keough, K. M. W. & Kariel, N. (1987) Biochim. Biophys. Acta 902, 11-18
25. Yeagle, P. L. (1989) The Biology of Cholesterol. CRC Press, Boca Raton, FL
26. Green, C. (1977) Int. Rev. Biochem. 14, 101-152
27. Schroeder, F., Jefferson, J. R., Kier, A. B., Knittel, J., Scallen, T. J., Wood, W. G. & Hapala, I. (1991) Proc. Soc. Expt. Biol. Med. 196, 235-252
28. Chapman, D. & Wallach, D. F. H. (1965) in Biological Membranes, vol. 1, Physical Fact and Function (Chapman, D., ed.), pp. 125-202, Academic Press, London
29. Davis, P. J. & Keough, K. M. W. (1983) Biochemistry 22, 6334-6340
30. Keough, K. M. W., Giffin, B. & Matthews, P. L. J. (1989) Biochim. Biophys. Acta 983, 51-55
31. Demel, R. A., Geurts van Kessel, W. S. M. & van Deenen, L. L. M. (1972) Biochim. Biophys. Acta 266, 26-40
32. Ghosh, D., Williams, M. A. & Tinoco, J. (1973) Biochim. Biophys. Acta 291, 351-362
33. Kariel, N., Davidson, E. & Keough, K. M. W. (1991) Biochim. Biophys. Acta 1062, 70-76
34. Van Blitterswijk, W. J., van der Meer, B. W. & Hilkmann, H. (1987) Biochemistry 26, 1746-1756
35. Litman, B. J., Lewis, E. N. & Levin, I. W. (1991) Biophys. J. 57, 273a
36. East, J. M., Jones, D. T, Simmonds, A. C. & Lee, A. G. (1984) J. Biol. Chem. 259, 8070-8071
37. Mitchell, D. C., Straume, M., Miller, J. L. & Litman, B. J. (1990) Biochemistry 29, 9143-9149
38. Jaworsky, M. & Mendelsohn, R. (1985) Biochemistry 24, 3422-3428
39. Anderle, G. & Mendelsohn, R. (1986) Biochemistry 25, 2174-2179
40. Marai, L. & Kuksis, A. (1973) Can. J. Biochem. 51, 1365-1379
41. Hidalgo, C., Ikemoto, N. & Gergely, J. (1976) J. Biol. Chem. 251, 4224-4232
42. Vrbjar, N., Simatos, G. A. & Keough, K. M. W. (1990) Biochim. Biophys. Acta 1030, 94-100
43. Vrbjar, N., Kean, K. T., Szabo, A., Senak, L. & Mendelsohn, R. (1992) Biochim. Biophys. Acta, 1107, 1-11
44. Mabrey, S. & Sturtevant, J. M. (1976) Proc. Natl. Acad. Sci. U.S.A. 73, 3862-3866
45. Davis, P. J., Fleming, B. D., Coolbear, K. P. & Keough, K. M. W. (1981) Biochemistry 20, 3633-3636
46. Coolbear, K. P., Berde, C. B. & Keough, K. M. W. (1983) Biochemistry 22, 1466-1473
47. Keough, K. M. W., Giffin, B. & Kariel, N. (1987) Biochim. Biophys. Acta 902, 1-10

Modulation of membrane lipid phase behaviour by chemical modification *in situ*

P. J. Quinn

Division of Life Sciences, King's College London, Campden Hill, London W8 7AH, U.K.

Introduction

The idea that chemical modification of membrane lipids could be achieved *in situ* was first demonstrated by Chapman and Quinn in 1976 [1]. The rationale underlying the work was that, if the unsaturated double bonds were largely responsible for the fluid character of the membrane lipid matrix, their saturation would result in a reduction in fluidity. Although simple in concept the practice required application of an entirely novel approach to the catalytic hydrogenation of lipids. It was found that conventional hydrogenation catalysts, such as Adam's catalyst, were unable to bring about hydrogenation of unsaturated lipids dispersed in aqueous systems. The solution to this problem was to employ homogeneous catalysts in which complexes of transition metal atoms were linked to suitable ligands that are able to gain access to the lipid substrate arranged in a bilayer configuration. The initial work was performed using rhodium complexes with triphenyl phosphines designed for hydrogenation and hydroformylation reactions in organic solvents [2] but, subsequently, water-soluble homogeneous catalysts [3] were found to be active against lipid substrates in aqueous dispersions.

One of the objectives of performing hydrogenation reactions *in situ* was to modulate the fluidity of biomembranes and to examine the role of membrane lipid fluidity in biochemical and physiological functions. Additionally, because membrane lipids with six or more unsaturated double bonds were found to be significant components of some membranes, such as the retinal rod membranes of the eye, the hydrogenation of these lipids was thought to be a useful tool for identifying their role in these membranes.

Once the idea of chemical modification with homogeneous catalysts of membrane lipids had been demonstrated it was not long before other types of reaction were contemplated. The first was polymerization. Initial attempts to polymerize lipids of natural origin containing unsaturated double bonds in the presence of a polymerizing agent were singularly unsuccessful, but again it was Chapman's group that was amongst the first to demonstrate that synthetic lipids with acetylenic groups could, under appropriate conditions, react to form polymerized films [4]. Another possibility involved combining the highly ordered arrangement of phospholipids in a bilayer with the potential selectivity of homogeneous catalysts to perform highly specific reactions. The first demonstration of such reactions was the successful discrimination of hydrogenation and hydroformylation of terminal olefins differing in chain length by only one carbon atom [5].

The aim of this chapter is to review developments since the introduction of these techniques, with particular emphasis on the use of homogeneous catalysts in hydrogenating model and biomembrane systems and the information that has been gained about the role of lipid fluidity in membrane functions. The use of polymerizable acetylenic lipids in model and biological structures will also be examined.

Hydrogenation of unsaturated membrane lipids

When unsaturated membrane lipids are dispersed in aqueous media they aggregate into structures that are inaccessible to conventional heterogeneous catalysts. Likewise, lipids of biological membranes arranged in bilayer configuration cannot be hydrogenated in the presence of such catalysts. Atomic complexes of transition metals such as nickel, copper, platinum, palladium and ruthenium, have been shown to be active in hydrogenation reactions in homogeneous reaction systems. Most transition metals can be formed into complexes with hydrogenation activity; however, some complexes are considerably more active than others. Furthermore, the physical properties of the catalyst can be tailored to achieve highly selective hydrogenation reactions by either influencing the interaction of catalyst with the substrate or by varying solubility in the reaction medium. The catalytic complexes consist of atomic forms of the transition metal, ensuring that all the metal atoms of the catalyst can participate in reactions, thereby making the reaction more efficient in terms of the amount of catalyst required to sustain a given rate of hydrogenation.

Homogeneous hydrogenation catalysts

Many transition metal complexes capable of activating molecular hydrogen are known [6]. Most of these complexes have been shown to catalyse the efficient reduction of unsaturated bonds, including olefinic $-C{=}C-$, $={C}{=}O$ and $={C}{=}N-$. When using such catalysts in biological systems, however, there are a number of factors that need to be taken into account. In the case of living organisms, for example, it is essential that the catalyst is non-toxic or at least that the level of toxicity at concentrations required to sustain a reasonable level of hydrogenation is low. Toxicity can arise by breakdown of the catalyst complex and liberation of the transition metal element and/or ligands of the complex, either of which may be toxic. Furthermore, side reactions other than hydrogenation may lead to the formation of unwanted, although not necessarily toxic, byproducts. Such reactions include ligand exchange with biomolecules resulting in complexes with altered catalytic properties. Side reactions are potentially damaging in the case of sulphonated derivatives of Wilkinson's catalyst, for example, where the catalytically active species, $RhH(SP\phi_2)_3$ and $[Rh(SP\phi_2)_3]^+$, are known to hydrogenate =C=O functions in addition to *cis* unsaturated bonds of olefins. Reaction of biochemical compounds of a susceptible chemical configuration could have repercussions for cell viability.

Chemical catalysis is often performed under conditions of temperature etc. that are well outside the physiological range. In biological applications, the catalyst complex must be stable under the conditions required to preserve stability of biomembranes or viability of living organisms. At the same time reasonable reaction rates must be sustained under these physiological conditions. Ideally the presence of the catalyst in the system should not affect any properties of the membrane other than its response to the altered level of saturation of the constituent lipids. This can be achieved by removal of the catalyst complex at the completion of the hydrogenation reaction.

Because of these relatively stringent requirements there are comparatively few complexes that are suitable for biological applications. The group of complexes such as $[Co(CN)_5]_3^-$ for example, although very active under conditions appropriate for hydrogenation of biological membranes, are stable only in the presence of excess cyanide [7]. Another common ligand, 2-aminopyridine, in catalysts such as $RuCl_2(2-Ampy)_2$, although producing highly active catalysts under relatively mild conditions, is highly toxic to living cells. Another group of the type $RuCl_n(H_2O)_{6-n}$ requires high temperatures and concentrated chloride solutions to produce even modest rates of hydrogenation of unsaturated fatty acids [8]. Finally, the classic group of organometallic compounds containing low-valent transition metal ions are largely unsuitable because of their unstable character in aqueous media.

Water-insoluble homogeneous catalysts
The first catalyst used in homogeneous catalytic hydrogenation of membrane lipids was Wilkinson's catalyst [9, 10]. The chemistry of the hydrogenation process has been described in detail for this and related catalysts [2]. The mechanism of catalytic hydrogenation of alkenes involves three steps: activation of molecular hydrogen; activation of the substrate; and transfer of hydrogen to the substrate to form the saturated product. When Wilkinson's catalyst is dissolved in the presence of hydrogen, one of the triphenylphos-

phine groups is displaced and hydrogen adds oxidatively to the rhodium to give a 5-co-ordinate rhodium (III) complex in a reaction summarized as follows:

$$[Rh(PPh_3)_3Cl] + H_2 \leftrightarrow [Rh(PPh_3)_2H_2Cl] + PPh_3$$

As shown, the reaction is freely reversible, so that when hydrogen is removed, the pale yellow solution typical of the hydrogenated form of the catalyst darkens again to the original orange colour.

On addition of an alkene, this co-ordinates to the potential vacant site on the dihydride complex as follows:

$$[Rh(PPh_3)_2H_2Cl] + RCH=CH_2 \leftrightarrow [Rh(PPh_3)_2H_2Cl(RCH=CH_2)]$$

The final stage of the hydrogenation involves hydrogen transfer to the co-ordinated alkene, most likely in a two-step process, to give an alkane. The catalytic cycle is completed when the complex takes up a further molecule of hydrogen and alkene to form an activated ternary complex.

The metal catalyst therefore cleaves the H—H bond of molecular hydrogen (homolytic splitting of the hydrogen molecule), weakens the C=C bond through the formation of a coordination complex, and brings the two hydrogen atoms and alkene into sufficiently close proximity to enable the transfer reaction to occur at an efficient rate.

One of the features of Wilkinson's catalyst that limits its use in biomembrane systems is its low solubility in water. Nevertheless, the catalyst is active against unsaturated lipid substrates, although at a somewhat reduced rate. This can be seen in Fig. 1 which shows the relationship between the initial rate of hydrogenation of soya phosphatidylcholine in aqueous mixtures of tetrahydrofuran in the presence of Wilkinson's catalyst. It can be seen that the initial rate of hydrogen uptake decreases as the proportion of water in the system increases, and reaches a limiting rate when the phospholipid assumes a bilayer form. Despite differences in initial reaction rate, virtually complete hydrogenation occurs in all combinations of solvent. In adapting water insoluble catalysts for use in biological systems it is necessary to introduce the catalyst into the membrane using a solvent vector. Solvents such as tetrahydrofuran and dimethylsulphoxide have been found to be useful. The introduction of catalyst in a minimum amount of solvent, which is miscible with water, causes the insoluble complex to partition into the hydrophobic domain created by the lipid substrate. The catalyst obviously cannot be removed subsequently from the substrate without destroying the integrity of the membrane. It is also important to verify that the solvent used to introduce the catalyst does not perturb the stability of the membrane. The original studies of hydrogenation of phospholipids dispersed in aqueous systems were performed using Wilkinson's catalyst introduced in a solvent vector of tetrahydrofuran [1, 11, 12]. It was shown that complete hydrogenation of the dispersed lipid could be achieved under relatively mild conditions of temperature, hydrogen pressure and catalyst concentration. Biophysical studies employing differential scanning calorimetry and X-ray diffraction confirmed that the solvent used to deliver the catalyst and the presence of catalyst in the lipid bilayer did not drastically alter the structural properties of the membrane. Wilkinson's catalyst has also been used to hydrogenate model

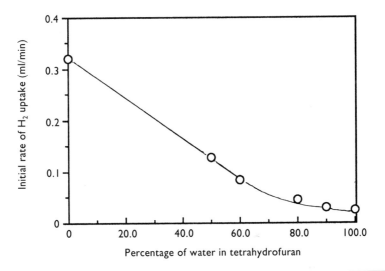

Fig. 1. Initial rate of uptake of hydrogen by soya phosphatidylcholine dispersions in aqueous tetrahydrofuran in the presence of Wilkinson's catalyst

(Data from [1].)

membranes prepared from lipid extracts of rat liver mitochondria and microsomes and human erythrocytes [13].

It was found that the presence of cholesterol markedly influences the hydrogenation of mixed phospholipid dispersions [14]. This could be explained by the fact that cholesterol restricts partition of the catalyst into the lipid bilayer structure. It was also found that no dihydrocholesterol forms during hydrogenation of the phospholipid, showing that cholesterol is not a substrate for reaction under the conditions employed.

Water-soluble homogeneous catalysts
A major advance in the application of homogeneous catalytic hydrogenation methods to the modulation of lipid phase behaviour was the use of water-soluble catalysts. The need to employ solvent vectors to introduce the catalyst into the membrane can be avoided and there is more scope for removal of the catalyst at the end of the reaction. Water-soluble complexes can be

removed simply by washing, gel filtration, density gradient centrifugation and, in the case of charged complexes, by adsorption to ion-exchange resins.

Synthesis of the first water-soluble catalyst complexes was reported by Joo and Beck [3] and involved the replacement of triphenylphosphine with sulphonated triphenylphosphine. The sulphonated derivative was found to stabilize the lower oxidation states of a number of transition metals, such as Rh, Ru, Ir, Pt, Ni and Cu, in aqueous systems and these water-soluble catalysts facilitated hydrogenation of soluble substrates such as pyruvic acid.

The water-soluble complexes appear to have very similar chemical properties to their non-sulphonated triphenylphosphine counterparts. The presence of the charged sulphonyl group renders the catalyst complex very soluble in neutral aqueous solutions [15], and solubility can be modulated by salt concentration or pH. The sulphonate group is not generally coordinated to the metal, and infrared spectra indicate only minor differences in electronic state of the central metal ion compared with the triphenylphosphine complexes. The synthesis and reactivity of a range of monosulphonated triphenylphosphine complexes have been reported [16, 17]. The solubility of metal complexes of phosphines in water can be increased by using multisulphonated triphenylphosphine [18, 19]. These types of catalysts do not penetrate readily into lipid substrates, but partition can be influenced by the use of complexes with phosphine-like ligands [20] or by attachment of amphiphilic, long-chain aliphatic ternary phosphines to the metal [21].

Homogeneous catalytic complexes containing triphenylphosphine ligands are generally unstable in the presence of oxygen and this places a major limitation on their use with living organisms under aerobic conditions. This problem has been largely overcome by synthesis of catalytic complexes based on sulphonated alizarine derivatives of Ru and Pd [22]. The Pd (II) alizarine complex is not only resistant to inactivation by oxygen, which renders it more stable over relatively long reaction times, but also readily soluble in water [23]. As it retains high activity under physiological conditions it need only be added to biological systems in trace amounts, thereby avoiding toxicity problems. Toxicity arises not only from the metal ions and ligands but also from the detergent action of these surface-active complexes.

Hydrogenation of unsaturated phospholipids dispersed in aqueous systems using a water-soluble homogeneous catalyst was first reported by Madden and Quinn [24]. The catalyst was a sulphonated derivative of Wilkinson's catalyst which did not appear to affect the structure of bilayers with respect to their permeability barrier properties [25]. The catalyst was found to hydrogenate oil-in-water emulsions and two-phase oil–water systems without the need for organic co-solvents [26]. The reaction rate could be increased significantly by screening the electrostatic charge on the sulphonate groups with inorganic cations added to the aqueous phase. This allowed the catalyst to penetrate into the substrate at the interface; partition of the catalyst from the aqueous phase into the lipid phase could not be detected. Further evidence for exclusion of catalyst from the lipid phase can

Fig. 2. Hydrogenation of multilamellar dispersion (●) and single bilayer vesicles (○) of soya lecithin in the presence of the sulphonated derivative of Wilkinson's catalyst

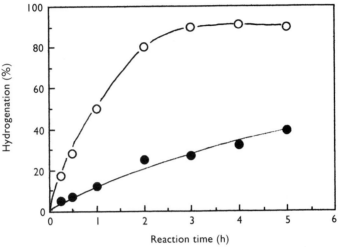

(Data from [24].)

be seen in the comparison of hydrogenation rate of multibilayer dispersions of unsaturated phospholipids and highly dispersed vesicular suspensions, as illustrated in Fig. 2. The reaction rate in multilamellar dispersions could be accelerated by dispersing the phospholipid in the presence of the catalyst rather than as shown in Fig. 2 where catalyst was added to dispersed substrate.

The reactivity of the water-soluble paladium catalyst, palladium di(sodium) alizarine monosulphonate [Pd(QS)$_2$], has been examined in multilamellar dispersions of unsaturated phospholipids [27]. With substrates of dioleoylphosphatidylcholine there is a transient appearance of *trans* ω—9, but no *cis* double bonds were observed when the *trans* ω—9 derivative of phosphatidylcholine was used as substrate. It was suggested that hydrogenation may proceed by a *cis-trans* isomerization followed by reduction of the *trans* double bond. Hydrogenation of di—18:2 and di—18:3 derivatives of phosphatidylcholines show highly complex patterns of partially saturated molecular species including combinations of *cis* and *trans* positional isomers with little evidence of bond migration. Comparisons of the rate of hydrogenation of unsaturated molecular species of phosphatidylcholines with those of dioleoylphosphatidylethanolamine revealed that the reduced Pd(QS)$_2$ catalyst had a slight preference for phosphatidylcholines. A preference for polyunsaturated molecular species compared with the monounsaturated molecular species of phosphatidylcholine was also observed. Differences in the accessibility of catalyst to substrates presented in bilayer form compared with those in hexagonal-II configuration may explain the different susceptibilities exhibited by phosphatidylcholines and phosphatidylethanolamines. These differences persisted in mixed dispersions hydrogenated at temperatures at which phase separations of bilayer and hexagonal-II structure would be expected to occur in the substrate.

Hydrogenation in the presence of homogeneous catalysts is conventionally performed in the presence of hydrogen gas. The rate of reaction can be increased at physiological temperatures by increasing the pressure of hydrogen. Although many biological systems can be preserved under these conditions, hydrogen gas is not the most convenient form of hydrogen and alternative strategies have been explored. Several classes of compounds including amines, alcohols, sugars, silicon or tin hydrides, can serve as hydrogen donors in catalysed hydrogen transfer but, in general, the conditions required for a meaningful conversion are not biocompatible. The photochemical reaction of the ruthenium bipyridyl complex [$RuCl_2(bipy)_3$] together with ascorbic acid as a sacrificial electron donor [27, 28], has been shown to catalyse the reduction of water and generate molecular hydrogen in the presence of water-soluble Wilkinson's catalyst [29]. This system has been exploited for catalysing the light-dependent hydrogenation of phospholipid multilayer dispersions, emulsified triacylglycerols and membranes of the living protozoan *Tetrahymena pyriformis* [30].

Hydrogenation of biological membranes

We have seen that homogeneous catalyst complexes are active against unsaturated phospholipids dispersed in aqueous systems and under physiological conditions of temperature and electrolyte concentration. Although anaerobic conditions in the presence of hydrogen gas are required for hydrogenation, biological membranes and cells can tolerate such conditions, often for relatively long periods, without deleterious effects. The effects of hydrogenation of isolated biological membranes, suspensions of subcellular organelles and preparations of living cells have all been reported.

Chapman and Quinn [1] first reported the successful hydrogenation of biological membranes. Partial hydrogenation of microsomal suspensions prepared from liver and muscle was achieved in the presence of Wilkinson's catalyst under 101 325 Pa (1 atm) of hydrogen. Suspensions of intact mitochondria were also hydrogenated under these conditions. Hydrogenation of membrane lipids of a living organism was demonstrated shortly afterwards [30a]. Studies of the effects of hydrogenation on biological membranes will be described below.

Effect of lipid saturation on cation pumps
The effect of hydrogenation of lipids of rabbit hind leg muscle sarcoplasmic reticulum preparations on the activity of Ca^{2+}-ATPase pump protein has

Table I. The effect of hydrogenation of rabbit muscle sarcoplasmic reticulum on activity of Ca^{2+}-ATPase

Preparation	Ca^{2+}-ATPase (IU)	Hydrogenation (%)
Control (N_2)	2.65	0
Hydrogenated (H_2)	2.62	25

Suspensions of sarcoplasmic reticulum were incubated for 3 h at 20°C under 911 925 Pa (9 atm) gas pressure in the presence of 6 mol Wilkinson's catalyst per 100 mol membrane phospholipid.

been reported [31]. It was shown that up to 35% of the unsaturated bonds could be saturated during 5 h incubation at 911 925 Pa (9 atm) H_2 in the presence of Wilkinson's catalyst. ATPase activity was completely inhibited on adding catalyst, but this could be prevented by preserving the catalyst in its hydride form. Table 1 shows the effect of hydrogenation of sarcoplasmic reticulum on calcium pump activity when assayed in buffers saturated with H_2. It can be seen that removal of 25% of *cis* double bonds does not affect the activity of Ca^{2+}-ATPase. This result could be interpreted in any of the following ways: (i) removal of 25% of the unsaturated double bonds does not significantly decrease membrane fluidity; (ii) activity of the enzyme is not dependent on membrane fluidity; or (iii) only a selected group or pool of lipids are hydrogenated. Selective hydrogenation may arise from preference of the catalyst for polyunsaturated fatty acyl chains. No direct methods were employed to determine the effect of hydrogenation on membrane fluidity. It should also be noted that only the rate of ATP hydrolysis was measured; the translocation of calcium across the membrane or changes in the passive permeability of the membrane to calcium ions were not examined.

Membrane homeoviscous adaptation

Many organisms are known to adapt the extent of unsaturation of their constituent membrane lipids in response to environmental factors, most notably temperature [32, 33]. The question of whether the activity of acyl chain desaturase enzymes is altered directly by the change in environmental factors, or indirectly by a change in the local viscosity of the lipid domain in which the enzymes are integral components, has been examined by homogeneous catalytic hydrogenation.

One system that has been studied is that in microsomes prepared from potato tubers [34]. Oleoyl-CoA desaturase from potato tuber microsomes is believed to consist of four intrinsic membrane proteins [35, 36]: (i) a reductase that transfers electrons from NADH to (ii) an electron carrier; (iii) a lysophosphatidylcholine-acyltransferase; and (iv) oleoylphosphatidylcholine desaturase. Cytochrome b_5 is believed to be the electron carrier involved in oleate desaturation, and NADH–cytochrome b_5 reductase [37] and cytochrome b_5 [38] from potato tuber microsomal preparations have been isolated and characterized. When membrane lipids of microsomal suspensions were hydrogenated in the presence of $Pd(QS)_2$ catalyst there was a marked rigidification of the hydrocarbon domain of the membrane as judged by electron spin resonance probe measurements and a stimulation of NADH reductase using ferricyanide as the electron acceptor. It was suggested that the increased NADH–ferricyanide reductase activity was caused by greater accessibility of the active site of the enzyme to the electron acceptor, ferricyanide, possibly as a result of a displacement of the protein with respect to the membrane lipid matrix as the lipids become less fluid. Vertical displacement of proteins in these circumstances has also been suggested by other studies [39]. In contrast, when cytochrome c replaced ferricyanide as the electron acceptor it was found that NADH–cytochrome reductase as well as oleoyl residue acylation and desaturation were markedly inhibited by saturation of the membrane lipids. The loss in activity of these components, however, may have been caused by the catalyst having a direct effect on components of the electron transport chain rather than on the level of unsaturation of the membrane lipids. Another factor that also needs to be considered in experiments of desaturation is the risk of creating unusual molecular species of lipid which may block metabolic reactions such as desaturation by processes of competitive inhibition [40, 41].

Similar conclusions were drawn from studies on 18:1-CoA desaturase activity in yeast microsomal membranes hydrogenated in the presence of Pd-complex [42]. In this system there was a clear indication of a dependence of lipid desaturase activity on the lipid microviscosity as judged by fluorescence probe polarization measurements.

Membrane lipid biosynthetic pathways
The possibility of modifying the extent of saturation of membrane lipids at sites remote from their site of synthesis in living cells provides a tool for examining the factors regulating lipid homeostasis. Information on the pathways of unsaturated membrane lipid biosynthesis and processes of redistribution from the site of synthesis to the different subcellular membranes has been obtained from studies of the unicellular green alga, *Daniella salina*. This cell, in common with other plant cells, has a complex mechanism of membrane lipid biosynthesis. Some lipids are synthesized in the endoplasmic reticulum, others in the chloroplast and others partly in both locations. Membrane lipids of *Daniella* can be hydrogenated extensively in the presence of water-soluble palladium alizarine catalyst under conditions that permit full recovery of the cells within 24 h [43]. When cells are incubated with the catalyst under 101 325 Pa (1 atm) of hydrogen for less than 2 min, only the unsaturated lipids of the surface (plasma) membrane are reduced in number. Cells treated in this way cease growth for about 12 h during which time the hydrogenated acyl chains are partially reconverted to their original

level of unsaturation. Restoration of lipid unsaturation permits a resumption of growth as membrane functions are presumably restored. Subfractionation of hydrogenated cells showed that the plasma membrane component of the microsomal fraction was hydrogenated to the greatest extent and endoplasmic reticulum to a considerably lesser extent.

Attempts to hydrogenate the outer surface of the plasma membrane of *Tetrahymena mimbres* have also been reported [44] and >20% saturation has been achieved. There was, however, loss in viability of the cells when more than trace amounts of hydrogenation were detected. The cause of this sensitivity to hydrogenation of the plasma membrane is presently unknown.

Membrane lipid topology and function

The topology of lipids in the membranes of complex organisms or in subcellular membrane preparations can be probed by determining access to a hydrogenation catalyst. Water-soluble catalyst complexes, for example, are not readily permeable to membranes and, when added to suspensions of cells or closed vesicular structures, their action has been shown to be largely restricted to the outer monolayer, at least at short time intervals after commencement of the reaction (see [44]). Selective hydrogenation of lipid classes has also been observed. Analysis of the pattern of hydrogenation during incubation of pea chloroplasts in the presence of $Pd(QS)_2$ catalyst is a case in point [45]. As can be seen from Fig. 3, which shows the extent of hydrogenation of the three major galactolipid classes at intervals during the hydrogenation reaction, there is a marked difference in susceptibility of membrane lipids to hydrogenation, with galactolipids more readily hydrogenated than the acidic lipids of the membrane. This effect could result from a charge repulsion between the negatively charged functional groups of sulphoquinovosyldiacylglycerols and phosphatidylglycerols and the sulphonated alizarine groups on the palladium catalyst complex.

The role of unusual lipids in membrane structure and stability can also be investigated using the hydrogenation method. There has been considerable speculation, for example, on the role of phosphatidylglycerol containing *trans*-Δ^3-hexadecenoic acid in stabilizing the oligomeric form of the light-harvesting chlorophyll–protein complexes of the thylakoid membrane. Recently, the role of this unusual fatty acyl residue was investigated by Horvath *et al.* [46] using *D. salina*. They found that growth at 15°C compared with 30°C resulted in a significant decrease in the proportion of the *trans* fatty acyl molecular species of phosphatidylglycerol and there was a corresponding reduction in stability of the light-harvesting chlorophyll–protein complexes isolated from the membranes after subjection to gel electrophoresis. Catalytic hydrogenation was observed to take place at a slower rate with *trans* than with *cis* fatty acid residues, but its eventual reduction to low levels did not result in a reduction in oligomer formation of the chlorophyll complexes. This suggested that factors other than the decrease in *trans*-Δ^3-hexadecenoate molecular species associated with growth

Fig. 3. Hydrogenation *in situ* of individual lipid classes of pea thylakoid membranes in the presence of paladium alizarine catalyst as a function of reaction time

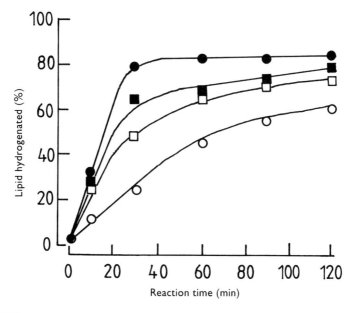

●, *monogalactosyldiacylglycerol*; ■, *digalactosyldiacylglycerol*; □, *sulphoquinovosyldiacylglycerol*; ○, *phosphatidylglycerol. (Data from [45].)*

at lower temperature were responsible for the decreased stability of the light-harvesting chlorophyll–protein complexes.

Adaptation of membranes to cold stress

Catalytic hydrogenation has been used to examine the mechanism of retailoring membrane lipids in the process of cold adaption in *Tetrahymena pyriformis* [46]. Isolated cilial membranes, when hydrogenated in the presence of Pd (II) sulphonated alizarine complex, showed a marked increase in order parameter and rotational correlation time of electron spin resonance probes as the constituent lipids became saturated. This was associated with a dramatic decrease in endogenous phospholipase A activity of the membrane even when only a small proportion of the unsaturated bonds had been hydrogenated. The way in which an endogenous phospholipase responds to the change in physical state of its substrate is believed to be the mechanism whereby the biochemical changes responsible for thermal adaptation are bought about. This has been confirmed by studies of endogenous phospholipase activity in rat liver mitochondria where it was found that polyunsaturated species of membrane lipids are required to provide access of phospholipase A to its substrate [41].

Some organisms or differing strains of the same organism suffer loss in viability resulting from a sudden exposure to cold. These effects appear to be related to the extent of unsaturation in lipids of constituent membranes. Catalytic hydrogenation has proved to be a useful method for investigating the molecular basis of chilling sensitivity. The phase behaviour of membrane lipids of the blue-green alga *Anacystis nidulans* is believed to underly the chilling sensitivity of this organism [47], and catalytic hydrogenation studies have been undertaken to examine this hypothesis [48, 49]. The initial phase of membrane lipid hydrogenation appears to be confined to the outer cell surface membranes but, subsequently, the catalyst becomes accessible to the inner thylakoid membrane. Hydrogenation of the outer cytoplasmic membrane was associated with an increased leakiness to solutes, and phase separation of the membrane components occurred at relatively high temperatures. Some of the effects of hydrogenation of whole cells are illustrated in Fig. 4. This shows that chilling sensitivity of 28°C-grown cells is considerably less than that of 38°C-grown cells, as judged by K^+ release from the cell, representing leakiness of the cytoplasmic membrane, and rate of O_2 evolution, an index of integrity of the thylakoid membrane. When the membranes of 28°C-grown cells are subjected to hydrogenation they show a similar susceptibility to chilling injury as 38°C-grown cells. This strongly supports the hypothesis that the phase behaviour, as modulated by the degree of unsaturation of the constituent membrane lipids, is directly related to the susceptibility of the cells to chilling damage.

Membrane stability at high temperatures

Adaptation of organisms to elevated temperature is often associated with a shift in the molecular species of membrane lipids to more saturated fatty acyl substituents. It is often argued that this change renders the membrane more stable at elevated temperatures. This hypothesis has been examined in considerable detail in chloroplast photosynthetic membranes which are ideal for hydrogenation studies because of the highly unsaturated lipids present and the dependence of the membrane on these lipids for maintaining stuctural stability and organization [50]. The original studies were performed using Wilkinson's catalyst [51] and it was found that decreases of up to 40% of unsaturated bonds did not alter the ultrastructural features of the membrane or photosynthetic electron transport processes. Later studies using water-soluble catalysts [52, 53] showed that saturation of the lipids results in a decrease in electron transfer between the 'primary' electron acceptor QA and 'secondary' acceptor, QB. Fluorescence-induction kinetics indicated that there is an optimal level of lipid unsaturation for maintaining an efficient electron transfer from QA^- to the plastoquinone pool. Furthermore, the proportion of photosystem-IIb, which has a reduced complement of light-harvesting chlorophyll-II [54–56], compared with photosystem-IIa, the form

Fig. 4. **Effect of hydrogenation of membrane lipids of *Anacystis nidulans* on (a) K⁺ release and (b) photosynthetic oxygen evolution when chilled to different temperatures**

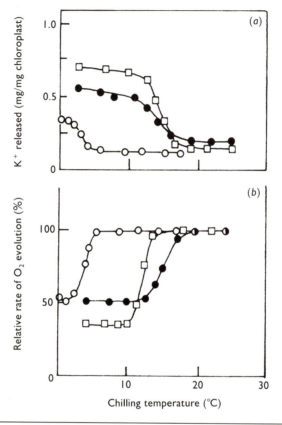

○, Cells grown at 28°C; □, cells grown at 38°C; ●, cells grown at 28°C and hydrogenated in the presence of palladium alizarine catalyst. (Data from [46].)

of photosystem-II with complete peripheral chlorophyll a/b light-harvesting chlorophyll-II, increases with increased hydrogenation of the membrane lipid.

Hydrogenation of the lipids of the photosynthetic membrane of higher plant chloroplasts prevents disruption of the membrane exposed to elevated temperatures. Freeze-fracture electron microscopic studies of hydrogenated chloroplasts has revealed the presence of particle-free domains within lipids [57]. When exposed to elevated temperatures (40–45°C), a treatment that results in a loss in photosynthetic electron transport [58], membrane destacking and dissociation of supramolecular protein complexes

[59], hydrogenated membranes show a reduced tendency to destack and vesiculate. Chlorophyll *a* fluorescence measurements and differential scanning calorimetry suggest that the hydrogenation of thylakoid membrane lipids causes an increased thermal stability of pigment protein complexes of the photosystem-II light harvesting apparatus. Similar stabilization of photosystem-I complex subjected to heat stress has also been observed [60].

Membrane fluidity and diffusion processes

Many membrane functions are known to require a fluid membrane as they rely on the diffusion and interaction between different membrane components. Most notable processes include electron transport chains of which fatty acyl-CoA desaturates have already been mentioned. Photosynthetic electron transport is another such system that has been examined using catalytic hydrogenation methods. Extensive hydrogenation of chloroplast suspensions has shown that inhibition of whole chain electron transport occurs before inhibition of either photosystem-II or photosystem-I activity is observed, suggesting that a fluid membrane is required to facilitate communication between the photosystems situated in different lateral domains in the membrane [61, 62].

To test the notion that plastoquinone diffusion between photosystem-II and cytochrome b_6-f complex was a rate-limiting step in photosynthetic electron transport when the fluidity of the thylakoid membrane was reduced, measurements were performed on the reduction rate of flash-oxidized cytochrome f in pea chloroplasts subjected to lipid hydrogenation [63]. The results of reduction of *cis*-unsaturated fatty acyl residues achieved in the presence of palladium alizarine catalyst, together with the effect on rotational correlation time of the spin-probe, 16-oxyl stearate, are shown in Table 2. It can be seen that a 30% reduction in unsaturated bonds results from the treatment with catalyst and there is a significant reduction in membrane fluidity as judged by the motion of the spin probe. Nevertheless, despite a reduction in the full-chain electron transport, no change in the rate of reduction of flash-oxidized cytochrome b_6-f was observed. This lead to the conclusion that the rate of diffusion of plastoquinol is unaffected by the reduced fluidity of the thylakoid membrane caused by the saturation of membrane lipids.

Hydrogenation of living cells

Hydrogenation of membrane lipids of a living organism was first described by Chapman *et al.* [30a], who accomplished the saturation of fatty acid residues of the membrane of *T. pyriformis* in the presence of Wilkinson's catalyst. The organism appeared to revive after partial hydrogenation of the membrane lipids but no long-term survival study was performed. Different cell types have been studied since then and, in general, cells have been found to vary considerably in their ability to survive hydrogenation of the plasma membrane. As noted above, the protozoan *T. nimbres* appears to be particu-

Table 2. Changes in fatty acyl composition of pea thylakoid membranes and motion of 16-deoxyl stearate resulting from hydrogenation in the presence of palladium alizarine catalyst

Fatty acid	Control	Hydrogenated
	Mole (%)	
16:0	11	12
16:1	2	2
16:3	1	trace
18:0	2	15
18:1	5	19
18:2	6	13
18:3	73	39
Rotational correlation time of 16-deoxyl stearate at 20°C	1.65s	2.07s

(Data from [63].)

larly sensitive [44]. Hydrogenation of lymphocytes with Wilkinson's catalyst has been reported [65] but this catalyst was found to be highly toxic to the cells. More success has been obtained with the use of the water-soluble palladium alizarine catalyst which has been used to hydrogenate plasma membranes of a living murine leukaemia cell line [66]. Survival of more than 80% of the cells was achieved under optimum treatment conditions which resulted in a 40% reduction of total cell fatty acid *cis* double bonds mainly in polyunsaturated 18:2, 20:4 and 22:6. Fluorescence probe measurements showed an increase of about 17-25% in structural ordering of the hydrophobic domain of the membrane compared with non-hydrogenated membranes. Another interesting consequence of hydrogenation of the plasma membrane was a change in expression of some cell-surface antigens but not others. Antigen expression was estimated by membrane immunofluorescence in a fluorescence-activated cell sorter employing monoclonal antibodies directed against specific cell-surface antigens followed by a sandwich labelling with fluorescent antibodies. It was shown that expression of a 15 kDa surface antigen (function unknown) was enhanced fourfold by hydrogenation whereas expression of the surface antigen designated H-2k and glycolipid antigen were unaffected by the treatment. It could not be established whether selective enhancement of expression of the 15 kDa antigen was caused by unmasking of cryptic sites or by some lateral rearrangement leading to more or less clustering of the antigens on the cell surface. In other experiments, the susceptibility of plasma membrane fractions and liposomes prepared from lipid extracts of membranes to hydrogenation in the presence of the palladium alizarine catalyst were compared and it was found that substrates that exhibited the highest fluidity were more readily hydrogenated. It was suggested that this resulted from a facilitated intercalation of the catalyst complex into the less ordered hydrocarbon domain of the bilayer structure.

Polymerization of lipids

The formation of polymers from crystalline diacetylenes was first described by Wegner [67]. This was followed by the polymerization of diacetylenic fatty acids [68]. Shortly after this, it was demonstrated that surfactant diacetylenes including phospholipid derivatives could polymerize at the air-water interface or as Langmuir-Blodgett multilayers [69, 70]. Since then, a variety of lipids bearing diacetylene, butadiene or vinyl groups in aqueous systems have been shown to form stabilized polymer structures [71-75].

The polymerization of diacetylenic lipids is a topotactic process as the propagation reaction is controlled by the position and orientation of successive monomers. Nevertheless, polymer formation does involve a bond rearrangement and a small overall conformation shift so that some flexibility in the structure is required. The perturbation associated with polymerization can be seen as a small change in the dimensions of the unit cell but the phase before and after polymerization is essentially crystalline [76]. Because the balance between a well ordered crystal structure and conformational flexibility for bond formation is finely poised, the extent of polymerization varies from virtually complete to highly unreactive. The diacetylenic derivatives of phosphatidylcholine fall generally into a moderately polymerizable category and typically liposomes of 1, 2-bis(tricosa-10, 12-diynoyl)-sn-glycerol-3-phosphocholine ($DC_{8,9}PC$), when irradiated with u.v. light at 254 nm, react to an extent of about 20%. No polymerization is observed at temperatures greater than the chain order-disorder phase transition and a close alignment of the chains in the ordered phase is required for extensive polymerization. The extent of polymerization can be significantly increased by incorporation of saturated short-chain spacer lipids into diacetylenic phospholipids, [77, 78] which tends to form an ideal mixture at high pressures [79]. An example of the enhancement of the polymerization of $DC_{8,9}PC$ in the presence of varying molar proportions of 1, 2-dinonanoyl-3-phosphocholine (DNPC) is shown in Fig. 5 where the fraction of $DC_{8,9}PC$ polymerized is plotted as a function of the molar fraction of DNPC. Optimal enhancement is achieved when the length of the short-chain spacer exactly matches that of the length of the saturated chain between the head group and the diacetylenic group. The spacer molecules are thought to allow interdigitation of the proximal region of the hydrocarbon chains or they may permit more flexibility in the central core of the bilayer while preserving the close packing of the proximal ends of the chains. Stuctural studies to date favour the latter interpretation [78].

Fig. 5. Fraction of $DC_{8,9}PC$ polymerized by 1 min u.v. irradiation (254 nm) when dispersed in water together with varying molar ratios of DNPC

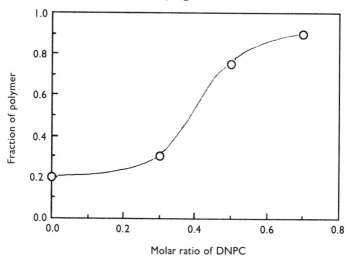

(Data from [78].)

The structure of diacetylenic phosphatidylcholines after polymerization has been examined by a range of biophysical methods. Fourier transform infrared spectroscopic studies of $DC_{8,9}PC$, one of the $DC_{m,n}PC$ series, and related phospholipids have been used to interpret changes in C=O stretching bands as representing distinct rotational isomers, implying a more ordered arrangement of the head group than corresponding bilayers of saturated phosphatidylcholines [80]. Electron density distribution maps constructed from X-ray scattering profiles of polymerized bilayers [81] indicate a lamellar repeat spacing that is relatively small and is comparable with those expected for saturated phosphatidylcholines with hydrocarbon chains 5–7 methylene carbons shorter. This introduced the notion that the hydrocarbon chains are tilted with respect to the plane of the bilayer in order to accommodate the additional space required for the longer acyl chains. Subsequent elastic neutron scattering experiments on these structures demonstrated that hydration of the head group was similar to that of saturated phospholipid [82].

The mesomorphic behaviour of diacetylenic phosphatidylcholine differs in many respects from their saturated counterparts, most notably by the formation of tubular structures in aqueous dispersions. Tubular structures of $DC_{8,9}PC$ can be formed in two ways. The first involves isothermal precipitation from ethanol/water mixtures [83]. The tubes (and helices) formed by this method have an average length of 150–300 μm and a diameter

(d) of 0.5–1 µm. The walls of the tubes can vary from 1–10 bilayer sheets stacked one upon the other. Alternatively, when large multilamellar vesicles in water are cooled through the chain disorder–order transition, shorter tubules are formed (30–60 µm) [84]. When the diacetylenic phospholipid is dispersed in the form of small unilamellar vesicles, the latter do not form tubules upon cooling through the acyl-chain phase transition, but instead they supercool to 2°C and transform, in an exothermic process, into large domains of flat bilayer or stacked bilayer sheets [85, 86]. Tubules can, however, be formed by heating the stacked bilayer sheets and subsequently cooling them down through the disorder–order transition.

The lamellar repeat spacing of diacetylenic phospholipids polymerized structure was found to be identical in both the tubular arrangement and bilayer sheets at low temperatures [87], with a value $d = 6.64$ nm. Wide-angle X-ray scattering indicates a crystalline-like packing of the chains in a fully extended all-*trans* configuration with a long-axis tilt of 32°. At higher temperatures the hydrocarbon chains undergo an endothermic transition associated with an increase in lamellar repeat spacing to $d = 7.4$ nm.

Protein reconstitution into polymerized lipids

The reconstitution of intrinsic membrane proteins into polymerizable lipids has been reported. In most of these studies the protein was incorporated into the polymerizable lipid membranes before irradiation with u.v. light to polymerize the bilayer. For example, purple membrane fragments have been incorporated into polymerized membranes by cosonication with diacetylenic lipid [88] and polymerizable sulpholipids [89]. In both cases the u.v. irradiation required to initiate the polymerization caused partial inactivation of the proteins. The ATP synthetase of *Rhodospirillum rubrum* has also been incorporated into polymerizable sulpholipid, but in this case polymerization results in a twofold stimulation in activity [90]. One method used to avoid these problems has been the insertion of proteins directly into polymerized bilayers and this has been achieved with both rhodopsin [91, 92] and bacteriorhodopsin [93]. In the latter case the high levels of incorporation achieved were attributed to mixing the polymerized lipid with short-chain phosphatidylcholine at a molar ratio of 2:1.

Polymerized liposomes as drug vectors

Entrapment or encapsulation of drugs and other agents in liposomes has been investigated extensively as a potential therapeutic delivery system. The possibility of using polymerized liposomes as a means of stabilizing these structures has been investigated by Chapman's group [94–96] and others [97]. One of the important features, as far as drug delivery applications are concerned, is that neither the polymerizable lipids themselves nor polymerized liposomes are non-thrombogenic [98].

Detailed studies of the permeability properties of polymerized liposomes have been reported [94]. It was found that diacetylenic phosphatidylcholine liposomes have a broad range of permeability characteristics. Permeability of liposomes, both before and after polymerization, depends on the structural nature of the diacetylenic lipid, i.e. whether it contains mixed or identical acyl chains, as well as the length and extent of polymerization of the lipid. Mixed-chain phosphatidylcholines, for example, were found to be more permeable after polymerization which was believed to be caused by disturbance of the close packing arrangement of the hydrocarbon chains. The stability of phospholipids with identical chains depended on the rate and extent of polymerization achieved.

Conclusions

The chemical modification of membranes containing unsaturated hydrocarbon substituents has been shown to be a useful tool in the study of the role of these lipids in membrane structure and stability. Homogeneous catalytic hydrogenation of biological membranes in isolated organelles or living cells has developed rapidly over the past few years with the introduction of more active catalytic complexes, especially under conditions of hydrogenation more compatible with living organisms. Advances in targeting catalysts to specific membranes and localizing action to specific membrane sites are likely to be important in future developments.

Polymerization of phospholipids, mainly using diacetylenic derivatives of natural phospholipids and particularly phosphatidylcholines, has been shown to produce highly stable and haemocompatible interfaces. Studies of these polymers to date show them to be unique materials that are likely to have wide applications in medicine, diagnostics and in the construction of molecular electronic devices.

The original impetus for the work owes much to the efforts of Dennis Chapman and later to the skill and dedication of Ference Joo and Laszlo Vigh in synthesis and application, respectively, of homogeneous catalysts.

References

1. Chapman, D. & Quinn, P. J. (1976) Proc. Natl. Acad. Sci. U.S.A. 73, 3971–3975
2. Osborn, J. A., Jardine, F. H., Young, G. F. & Wilkinson, G. (1966) J. Chem Soc. A71, 1711–1732
3. Joo, F. & Beck, M. T. (1973) Magy. Kem. Foly. 79, 189–191
4. Johnston, D. S., Sanghera, S., Pons, M. & Chapman, D. (1980) Biochim. Biophys. Acta 602, 57–59
5. Quinn, P. J. & Taylor, C. E. (1981) J. Mol. Catal. 13, 389–396
6. James, B. R. (1982) in Comprehensive Organometallic Chemistry (Wilkinson, G., Stone, F. G. A. & Abel, E. W., eds.), vol. 8, pp. 285–369, Pergamon Press, Oxford
7. James, B. R. (1973) Homogeneous Hydrogenation, Wiley, New York
8. Halper, J., Harrod, J. F. & James, B. R. (1966) J. Am. Chem. Soc. 88, 5150–5155
9. Young, J. F., Osborn, J. A., Jardine, F. H. & Wilkinson, G. (1965) Chem. Commun. 131–132
10. Quinn, P. J., Joo, F. & Vigh, L. (1989) Progr. Biophys. Mol. Biol. 53, 71–103

11. Chapman, D. & Quinn, P. J. (1976) Chem. Phys. Lipids 17, 363–372
12. Quinn, P. J., Chapman, D., Vigo, C. & Boar, B. R. (1977) Biochem. Soc. Trans. 5, 1132–1134
13. Vigo, C., Goni, F. M., Quinn, P. J. & Chapman, D. (1978) Biochim. Biophys. Acta 508, 1–14
14. Vigh, L., Joo, F., van Hasselt, P. R. & Kuiper, P. J. C. (1983) J. Mol. Catal. 22, 15–19
15. Salvesen, B. & Bjerrum, J. (1961) Acta Chem. Scand. 16, 735–741
16. Borowski, A. F., Cole-Hamilton, D. J. & Wilkinson, G. (1978) Nouv. J. Chim. 2, 137–144
17. Joo, F. & Toth, Z. (1980) J. Mol. Catal. 8, 369–383
18. Fontal, B., Orlewski, J., Santini, C. G. & Basset, J. M. (1986) Inorg. Chem. 25, 4320–4322
19. Kuntz, E. G. (1987) Chemtech. 17, 570–575
20. Wilson, M. E., Nuzzo, R. G. & Whitesides, G. M. (1978) J. Am. Chem. Soc. 100, 2269–2270
21. Farin, F., van Gaal, H. L. M., Bonting, S. L. & Daemen, F. J. M. (1982) Biochim. Biophys. Acta 711, 336–344
22. Bulatov, A. V., Nikitaev, A. T. & Khidekel, M. L. (1981) Izv. Akad. Nauk. SSSr Ser. Khim. 9, 924–925
23. Joo, F., Balogh, N., Horvath, L. I., Filep, G., Horvath, I. & Vigh, L. (1991) Anal. Biochem. 194, 34–40
24. Madden, T. D. & Quinn, P. J. (1978) Biochem. Soc. Trans. 6, 1345–1347
25. Madden, T. D., Peel, W. E., Quinn, P. J. & Chapman, D. (1980) J. Biochem. Biophys. Methods 2, 1–9
26. Drov, Y. & Manassen, J. (1977) J. Mol. Catal. 2, 219–222
27. Brown, G. M., Brunschwig, B. S., Creutz, C., Endicott, J. F. & Sutin, N. (1979) J. Am. Chem. Soc. 101, 1298–1300
28. Krishnan, C. V., Brunschwig, B. S., Crentz, C. & Sutin, N. (1985) J. Am. Chem. Soc. 107, 2005–2015
29. Oishi, S. (1987) J. Mol. Catal. 39, 225–232
30. Joo, F., Csuhai, E., Quinn, P. J. & Vigh, L. (1988) J. Mol. Catal. 49, 1–5
30a. Chapman, D., Peel, W. E. & Quinn, P. J. (1978) Ann. N.Y. Acad. Sci. 308, 67–84
31. Quinn, P. J., Gomez, R. & Madden, T. D. (1980) Biochem. Soc. Trans. 8, 38–40
32. Quinn, P. J. & Williams, W. P. (1985) in Photosynthetic Mechanisms and the Environment (Barber, J. & Baker, N. R., eds.), pp. 1–46, Elsevier, Amsterdam
33. Kates, M. (1990) in Glycolipids, Phosphoglycolipids and Sulfoglycolipids (Kates, M., ed.), pp. 235–320, Plenum Press, New York.
34. Demandre, C., Vigh, L., Justin, A. M., Jolliot, A., Wolf, C. & Mazliak, P. (1986) Plant Sci. 44, 13–21
35. Bonnerot, C. & Mazliak, P. (1984) Plant Sci. Lett. 35, 5–10
36. Gennity, J. M. & Stumpf, P. K. (1985) Arch. Biochem. Biophys. 239, 444–454
37. Galle, A. M., Bonnerot, C., Jolliot, A. & Kader, J. C. (1984) Biochem. Biophys. Res. Commun. 122, 1201–1205
38. Bonnerot, C., Galle, A. M., Jolliot, A. & Kader, J. C. (1985) Biochem. J. 226, 311–334
39. Shinitzky, M. (1984) Biochim. Biophys. Acta 738, 251–261
40. Schlame, M., Horvath, L. & Vigh, L. (1990) Biochem. J. 265, 79–85
41. Schlame, M., Horvath, J., Torok, Z., Horvath, L. & Vigh, L. (1990) Biochim. Biophys. Acta 1045, 1–8
42. Horvath, I., Torok, Z., Vigh, L. & Kates, M. (1991) Biochim. Biophys. Acta 1085, 126–130
43. Vigh, L., Horvath, I. & Thompson, G. A. (1988) Biochim. Biophys. Acta 937, 42–50
44. Pak, Y., Joo, F., Vigh, L., Katho, A. & Thompson, G. A. (1990) Biochim. Biophys. Acta 1023, 230–238
45. Horvath, I., Mansourian, A. R., Vigh, L., Thomas, P. G., Joo, F. & Quinn, P. J. (1986) Chem. Phys. Lipids 39, 251–264
46. Horvath, I., Vigh, L., Pali, T. & Thompson, G. A. (1989) Biochim. Biophys. Acta 1002, 409–412
47. Minorsky, P. V. (1985) Plant Cell Environ. 8, 75–94
48. Vigh, L. & Joo, F. (1983) FEBS Lett. 162, 423–427
49. Vigh, L., Joo, F., Droppa, M., Horvath, L. A. & Horvath, G. (1985) Eur. J. Biochem 147, 477–481
50. Quinn, P. J. & Williams, W. P. (1983) Biochim. Biophys. Acta 737, 233–266
51. Restall, C. J., Williams, W. P., Percival, M. P., Quinn, P. J. & Chapman, D. (1979) Biochim. Biophys. Acta 555, 119–130
52. Hileg, E., Rozsa, Z., Vass, I., Vigh, L. & Horvath, G. (1986) Photochem. Photobiophys. 12, 221–230
53. Horvath, G., Melis, A., Hideg, E., Droppa, M. & Vigh, L. (1987) Biochim. Biophys. Acta 891, 68–74

54. Melis, A. & Anderson, J. M. (1983) Biochim. Biophys. Acta 724, 473–484
55. Lam, E., Baltimore, B., Ortiz, W., Melis, A. & Malkin, R. (1983) Biochim. Biophys. Acta 724, 201–211
56. Melis, A. (1985) Biochim, Biophys. Acta 808, 334–342
57. Thomas, P. G., Dominy, P. J., Vigh, L., Mansourian, A. R., Quinn, P. J. & Williams, W. P. (1986) Biochim. Biophys. Acta 849, 131–140
58. Gounaris, K., Mannock, D. A., Sen, A., Brain, A. P. R., Williams, W. P. & Quinn, P. J. (1983) Biochim. Biophys. Acta 732, 229–242
59. Gounaris, K., Brain, A. P. R., Quinn, P. J. & Williams, W. P. (1984) Biochim. Biophys. Acta 766, 198–208
60. Vigh, L., Gombos, Z., Horvath, I. & Joo, F. (1989) Biochim. Biophys. Acta 979, 361–364
61. Vigh, L., Joo, F., Droppa, M., Horvath, L. A. & Horvath, G. (1985) Eur. J. Biochem. 147, 477–481
62. Horvath, G., Droppa, M., Szito, T., Mustardy, L. A., Horvath, L. I. & Vigh, L. (1986) Biochim. Biophys. Acta 849, 325–336
63. Gombos, Z., Barabas, K., Joo, F. & Vigh, L. (1988) Plant Physiol. 86, 335–337.
64. Reference deleted.
65. Peel, W. E. & Thompson, A. E. R. (1983) Leuk. Res. 7, 193–204
66. Benko, S., Hilkmann, H., Vigh, L. & van Blitterswijk, W. J. (1987) Biochim. Biophys. Acta 896, 1229–1235
67. Wegner, G. (1972) Makromol. Chem. 154, 35–46
68. Tieke, D., Wegner, G., Naegele, D. & Ringsdorf, H. (1976) Angew. Chem. 15, 764–765
69. Johnston, D. S., Sanghera, S., Manjon-Rubio, A. & Chapman, D. (1980) Biochim. Biophys. Acta 602, 213–216
70. Teike, D., Lieser, G. & Wegner, G. (1979) J. Polym. Sci. Polym. Chem. Ed. 17, 1631–1644
71. Day, D., Hub, H. H. & Ringsdorf, H. (1979) Isr. J. Chem. 18, 325–329
72. Hub, H. H., Hupfer, B., Koch, H. & Ringsdorf, H. (1980) Angew. Chem. 19, 938–940
73. Akimoto, A., Dorn, K., Gros, L., Ringsdorf, H. & Schupp, H. (1981) Angew. Chem. 20, 90–91
74. Hub, H. H., Hupfer, B., Roch, H. & Ringsdorf, H. (1981) J. Macromol. Sci. Chem. A 15, 701–715
75. Koch, H. & Ringsdorf, H. (1981) Makromol. Chem. 182, 255–259
76. Enkelmann, V. (1984) in Polydiacetylenes (Cantow, H. J., ed.), pp. 91–136, Springer-Verlag, Berlin
77. Singh, A. & Gaber, B. (1988) in Applied Bioactive Polymeric Materials (Gebelein, C. G., Carraher, C. E. & Forster, V. R., eds.), pp. 239–249, Plenum Press, New York
78. Rhodes, D. G. & Singh, A. (1991) Chem. Phys. Lipids 59, 215–224
79. Rhodes, D. G. & Poole, S. (1991) Biophys. J. 59, 636a
80. Rudolph, A. & Burke, T. (1987) Biochim. Biophys. Acta 902, 349–359
81. Rhodes, D. G., Blechner, S. L., Yager, P. & Schoen, P. (1989) Chem. Phys. Lipids 49, 39–47
82. Blechner, S. L., Skita, V. & Rhodes, D. G. (1990) Biochim. Biophys. Acta 1022, 291–295
83. Georger, J. H., Singh, A., Price, R., Schnur, J. M., Yager, P. & Schoen, P. E. (1987) J. Am. Chem. Soc. 109, 6169–6175
84. Yager, P. & Schoen, P. (1984) Mol. Cryst. Liq. Cryst. 106, 371–381
85. Burke, T. G., Sheridan, J. P., Singh, A. & Schoen, P. (1986) Biophys. J. 49, 321a
86. Burke, T. G., Rudolph, A. S., Price, R. R., Sheridan, J. P., Dalziel, A. W. & Singh, A. (1988) Chem. Phys. Lipids 48, 215–230
87. Caffrey, M., Hogan, J. & Rudolph, A. S. (1991) Biochemistry 30, 2134–2146
88. Pabst, R., Ringsdorf, H., Koch, H. & Dose, K. (1983) FEBS Lett. 154, 5–9
89. Yager, P. (1986) Biosensors 2, 363–373
90. Wagner, N., Dose, K., Koch, H. & Ringsdorf, H. (1981) FEBS Lett. 132, 313–318
91. Tyminski, P. N., Latimer, L. H. & O'Brian, D. F. (1985) J. Am. Chem. Soc. 107, 7769–7770
92. Tyminaki, P. N., Latimer, L. H. & O'Brian, D. F. (1988) Biochemistry 27, 2696–2705
93. Ahl, P. L., Smuda, R. P., Gaber, B. P. & Singh, A. (1990) Biochim. Biophys. Acta 1028, 141–153
94. Freeman, F. J., Hayward, J. A. & Chapman, D. (1987) Biochim. Biophys. Acta 924, 341–351
95. Leaver, J., Alonso, A., Durrani, A. A. & Chapman, D. (1983) Biochim. Biophys. Acta 732, 210–218
96. Johnston, D. S., McLean, L. R., Whittam, M. A., Clarke, A. D. & Chapman, D. (1983) Biochemistry 22, 3194–3202
97. Dorn, K., Klingbiel, R. T., Specht, D. P., Tyminski, P. N., Ringsdorf, H. & O'Brian, D. F. (1984) J. Am. Chem. Soc. 106, 1627–1633
98. Hayward, J. A., Levine, D. M., Nenfeld, L., Simon, S. R., Johnston, D. S. & Chapman, D. (1985) FEBS Lett 187, 261–266

Cubic and hexagonal lipid phases

Göran Lindblom and Leif Rilfors

Department of Physical Chemistry, University of Umeå,
S-90187 Umeå, Sweden

Introduction

In 1972 Singer and Nicolson [1] proposed their influential membrane model, where the lipids form a bilayer matrix, in which the working proteins are incorporated. A great deal of the experimental evidence supporting this model came from physicochemical studies, mainly n.m.r., e.s.r. and i.r. spectroscopic studies, X-ray diffraction and differential scanning calorimetry measurements, of which many were carried out by Chapman's laboratory at the time [2-11]. With the increasing number of physical investigations of biomembranes since then, our understanding of the role played by the lipids in the various membrane processes has changed. Thus, the lipids do not just form an inert and inactive thin film [1], but they participate in the function of the membrane in many ways. In particular, their ability to self-assemble into different aggregate structures is of great importance; i.e. the membrane lipids in water can form a variety of lyotropic liquid crystalline states as lamellar (L_α), normal hexagonal (H_I), reversed hexagonal (H_{II}) and cubic (I) phases (Fig. 1).

An increasing number of both experimental and theoretical works have been published over the past 5 years supporting the idea that the physicochemical properties of the lipids are a prerequisite for membranes to perform many of their functions. Most biological membranes contain at least one lipid that forms a non-lamellar phase by itself in water [12-16]. These include the following examples: phosphatidylethanolamine (PE) in the endoplasmic reticulum, Golgi complex, retinal rod membranes, and the plasma membrane of eukaryotes; PE and diphosphatidylglycerol (DPG) in the membranes of mitochondria; monogalactosyldiacylglycerol (MGalDG) in the membrane of chloroplasts; and PE, DPG and monoglucosyldiacylglycerol (MGlcDG) in the membranes of prokaryotes. Information has been obtained specifically about the structural importance and the functions of PE in model membranes, and some investigations have dealt with the influence of PE on bilayer permeability [17], membrane fusion [18] and orientational order and dynamics of the acyl chains [19–25]. Studies of the packing and functioning of the membrane lipids of the bacterium *Acholeplasma laidlawii* led to a major hypothesis on the regulation of the lipid composition in the membrane, based on the formation of non-lamellar phases ([13] and

Fig. 1. Structure of different liquid crystalline phases of membrane lipids

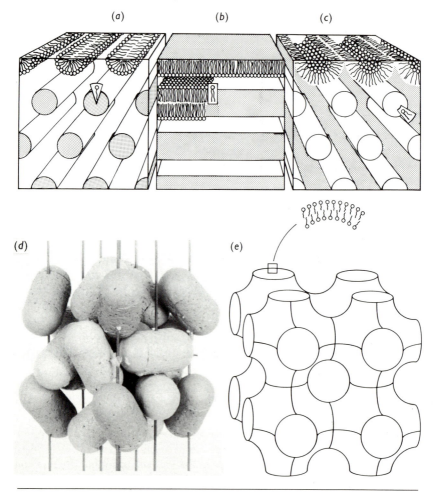

(a) Normal hexagonal (H_I) phase; (b) lamellar (L_α) phase; (c) reversed hexagonal (H_{II}) phase. (Figs. a–c reprinted from [12] with permission.) (d) Cubic phase built up of anisometric micelles. Reprinted from [114] with permission. (e) Bicontinuous cubic phase built up of lamellar units. The surface shown is an infinite periodic minimal surface that represents the midplane of the lamellar aggregates. Reprinted from [13] with permission.

references therein). A similar mechanism of lipid regulation has recently also been found to occur in other bacteria [26, 27].

Moreover, some membrane-bound enzyme activities have been shown to be lipid-dependent. Jensen and Schutzbach [28] reported that the activation of an enzyme that synthesizes dolichyl-P-mannose requires the presence of a lipid that forms non-lamellar structures, or the addition of hydrophobic peptides or alkanes that can induce a non-lamellar structure [16, 29].

The operational mechanism behind many membrane proteins and enzymes still has to be explored. An elucidation at the molecular level of such membrane processes requires studies of the interactions between lipids and proteins/peptides in the biological cell and/or in good model systems. Our working hypothesis is that membrane-bound proteins in some way can recognize the potential appearance of non-lamellar structures in the membrane and thereby take necessary measures to restore the original situation of the cell system. An understanding of the physical chemistry concerning the formation and structure of non-lamellar lipid phases is thus of fundamental biological relevance, irrespective of whether such phases, or their structural elements, actually form within the cell membranes or elsewhere in a biological system. Here, we will briefly review some of the physicochemical properties of membrane lipids forming non-lamellar phases.

Lipid-phase structures are influenced by various molecular properties

The different phases formed by membrane lipids in water are dependent on several factors: (i) the chemical structure and hydration properties of the lipid; (ii) the thermodynamic parameters such as temperature, hydrostatic and osmotic pressure, and pH; and (iii) the interaction with other molecules present in the aqueous region, in the hydrocarbon–water interfacial region, or in the hydrocarbon region of the lipid aggregates.

Chemical structure
The structure of the polar head-group is of great importance for the phase structures formed by lipid molecules. Under physiological conditions, lipids with small and/or less hydrated head-groups (PE, MGlcDG, MGalDG) often form H_{II} and I phases, while lipids with large and/or well hydrated head-groups [phosphatidylcholine (PC), phosphatidylserine (PS), phosphatidylglycerol (PG), DPG, phosphatidylinositol, diglucosyldiacylglycerol (DGlcDG), digalactosyldiacylglycerol] form liquid-crystalline lamellar (L_α) phases.

The structure of the hydrocarbon chains of the lipid also affects the phase structure. Generally, the temperature for the formation of a reversed non-lamellar phase is decreased when the chain length is increased, when double bonds or methyl branches are introduced into the chains, and when an ester linkage to the glycerol backbone is replaced by an ether linkage.

An important group of lipids in mammalian tissues is the sphingolipids, which have a sphingosine base, or a derivative of it, as a backbone. When glycerolipids and sphingolipids with identical or similar polar head-groups are compared, it can be concluded from the existing data that the sphingolipids have a greater capability to stabilize an L_α phase. Some anionic sphingolipids even form H_I phases and micellar solutions, and in this respect behave like ordinary soaps despite the fact that the sphingolipids contain two hydrocarbon chains [30].

Temperature, pressure and pH

The temperature often affects the phase equilibria of a membrane lipid/water system and, generally, a transition from an L_α phase to an I or H_{II} phase occurs at elevated temperatures. An increase in the hydrostatic pressure has a small, but reversed, effect.

Differences in hydration of the polar head-groups may play an important role, but the effects observed are rather complex. A transition from an L_α to an I or an H_{II} phase is often obtained when the water content is decreased in a lipid/water mixture. However, the phase sequence H_{II}-L_α-H_{II} was found by increasing the water content in PC/dodecane/water systems [29, 31]. In the systems mono-oleoylglycerol/water and MGlcDG/water [13], an increase in the water content leads to transitions from an L_α to an I phase, and from an I to an H_{II} phase, respectively.

The phase structure formed by lipids containing ionic head-groups can be affected by the pH value, so that at low pH values, a phase transition from L_α to H_{II} may take place. For the zwitterionic lipid, PE, the transition occurs when the pH is altered from alkaline to neutral, and for the anionic lipids, PS, DPG, and phosphatidic acid (PA), a change from a neutral to an acidic pH promotes such a transition. A striking observation, however, is that PE forms an H_{II} phase also when the pH is altered from a neutral to an acidic value.

Interaction with other molecules

Cations and other hydrophilic molecules are often able to induce phase transitions in membrane lipid/water systems. Divalent cations like Ca^{2+}, Mg^{2+} and Mn^{2+} induce an H_{II} phase from an L_α phase for the anionic lipids DPG and PA. Monovalent cations may promote the formation of an H_{II} phase; Li^+ has this effect on PS, and Na^+ affects PE and DPG, at molar concentrations. The fusogenic molecule polyethylene glycol has been reported to increase the temperature for the L_α-H_{II} phase transition (T_{LH}) of PE. Cryoprotectants such as trehalose influence T_{LH}, although conflicting results have been reported [32].

The influence of hydrocarbons such as n-alkanes, benzene and cyclohexane on the phase equilibria of above all PE, PC and PE-PC mixtures has been investigated ([29] and references therein). Generally, hydrocarbons promote or facilitate the formation of I and H_{II} phases. The ability of C_6–C_{20} n-alkanes to induce phase transitions decreases with increasing chain length of the alkane. With PC molecules, the water and dodecane concentrations required to form the non-lamellar phases increase with an increasing degree of acyl-chain saturation.

A number of amphiphilic molecules have been solubilized in membrane lipid bilayers, and their effect on the phase equilibria has been reported. The molecules investigated include alcohols, fatty acids, detergents, monoacylglycerols, diacylglycerols, sterols, anaesthetics, drugs, peptides and proteins. A discussion of the effects of all these molecules in various lipid/water systems would be too comprehensive here, and thus only examples will be given.

The shorter-chain n-alcohols (ethanol, butanol) stabilize the L_α phase, while the longer-chain n-alcohols promote the formation of an H_{II} phase in PE/water systems. Detergents and lyso-PCs form micellar solutions at high water contents, whereas an L_α phase might form when these lipids are mixed with lipids which by themselves form an H_{II} phase. Mono- and diacyl-glycerols are formed during fat digestion in the small intestine in vertebrates; both molecules induce the formation of I and H_{II} phases when mixed with PE or PC.

An overwhelming number of reports on the effect of cholesterol on the phase equilibria of lipid/water systems have been published. Usually, addition of up to approximately equimolar amounts of cholesterol to lipid will lead to the formation of I and H_{II} phases at a certain temperature. However, above these concentrations it has been found that the effect of cholesterol is reversed, leading to a stabilization of the L_α phase [33]. An L_α phase is formed when cholesterol is mixed in equimolar concentrations with Lyso-PC. The effect of cholesterol in some cases depends on the acyl-chain composition of the lipids. For example, for PE/PC mixtures, addition of the sterol promotes the formation of an H_{II} phase for PC with unsaturated acyl chains, but stabilizes the L_α phase for PC with saturated acyl chains. Moreover, cholesterol decreases the T_{LH} for PE with mono-unsaturated acyl chains, whereas the opposite effect is observed for human erythrocyte PE, having a large fraction of polyunsaturated hydrocarbon chains. Some steroids in which the hydroxyl group has been replaced by a keto group have a stronger ability than cholesterol to induce an H_{II} phase.

Recently, we reported a study of the effect of bile salts on the lipid-phase structure [34]. A partial-phase diagram of the ternary system dioleoyl-PE (DOPE)/sodium cholate/water was determined, and it was found that addition of even small amounts of cholate to the DOPE/water system leads to a transition from an H_{II} to an L_α phase. At higher cholate concentrations, an I phase (low water content) or a micellar solution phase (high water content) is present. It was concluded that cholate molecules have a strong tendency to alter the lipid monolayer curvature. This can be rationalized in terms of the molecular structure of cholate, which is amphipathic and has one hydrophobic and one hydrophilic side of the steroid ring system. The cholate molecules have a tendency to lie flat on the lipid aggregate surface [35, 36], thereby increasing the effective interfacial area of the polar headgroups, and altering the curvature free energy of the system.

Peptides and proteins constitute a very heterogeneous group of molecules regarding size, shape and charge. Contrary to molecules like alkanes, alcohols and detergents, it is therefore very difficult to generalize the effect of peptides and proteins on the phase equilibria of lipid/water systems. The hydrophobic peptide gramicidin has been studied extensively, and it has been found to induce I or H_{II} phases in PE/water and PC/water systems. The amphiphilic peptides cardiotoxin and mellitin interact both electrostatically and hydrophobically with lipid bilayers, and they form an H_{II} phase with DPG [37, 38].

Non-lamellar phases formed by two microbial glucolipids

MGlcDG has a pronounced ability to form reversed I and H_{II} phases besides the L_α phase. A phase diagram has been determined in the composition range 0.5–15 mol water per mol lipid at temperatures between -20 and 50 °C for dioleoyl-MGlcDG isolated from *A. laidlawii* [39]. Above 10 °C, only I and H_{II} phases exist; below 10 °C there is a region with an L_α phase. The I phase is bicontinuous and belongs to the space group *Ia3d*. The H_{II} phase stands in equilibrium with excess water, and the maximum hydration of this phase is about 11 mol water per mol lipid. Note that DOPE transforms from an L_α to an H_{II} phase at about 10 °C [40].

The lamellar–non-lamellar phase transitions for α- and β-anomers of MGlcDG containing straight, saturated acyl chains with 10–20 carbon atoms ($n = 10$–20) have been studied by McElhaney and colleagues [41–43]. The nature of this phase transition is affected by the length of the acyl chains and the anomeric linkage. The α-anomers with $n = 14$–16 form an I phase, while the compounds with $n = 17$–20 form an H_{II} phase; the temperature at which the non-lamellar phases form decreases from 105 °C for $n = 14$ to 77 °C for $n = 20$. The β-linked anomers are more prone to form non-lamellar phases. An I phase is formed when $n = 12$–15; the temperature for the formation of this phase increases from 58 °C for $n = 12$ to 73 °C for $n = 15$. An H_{II} phase is formed between 75 and 80 °C for the compounds with $n = 16$–20. Both the α- and β-anomers of MGlcDG with straight, saturated acyl chains are much more prone to form non-lamellar phases than the corresponding PEs [42].

The phase equilibria of the synthetic α-linked anomers of MGlcDG can be compared to those of MGlcDGs isolated from *A. laidlawii*. Three preparations of MGlcDG have been found to form an H_{II} phase at maximum hydration and physiological temperatures. The acyl chain composition of these preparations were as follows: (i) 97 molar % oleoyl chains (18:1c); (ii) 92 molar % elaidoyl chains; and (iii) 52 molar % palmitoyl chains (16:0) and 46 molar % 18:1c [39, 44]. However, an MGlcDG preparation containing 30 molar % 16:0, 47 molar % 18:1c, and 22 molar % short saturated acyl chains ($n = 12$–15) forms a mixture of lamellar gel (L_β) and L_α phases between 25 and 47 °C; above 47 °C a mixture of L_α and I phases is obtained, and a pure I phase is formed at about 70 °C [25]. Thus, when the average length of the saturated acyl chains in the third preparation above is shortened, MGlcDG no longer forms an H_{II} phase, but instead L_α and I phases. Moreover, from a freeze-fracture electron microscopy study of an MGlcDG preparation containing 72 molar % saturated acyl chains with an average length of 15.6 carbon atoms, and 24 molar % 18:1c, it was claimed that the lipid formed an L_α phase at 25 °C and an H_{II} phase at 55 °C [45]. These observations are in line with the results obtained from the synthetic α-anomers [42]. Consequently, it can be concluded that the ability of the α-anomer of MGlcDG to form H_{II} and I phases increases with an increasing length and degree of *cis* unsaturation of the acyl chains.

MGlcDG takes up smaller amounts of water than both PC and PE with corresponding acyl chains. The maximum hydration of MGlcDG at 35 °C is approximately 7, 8 and 11 mol water per mol lipid for dielaidoyl-, palmitoyl/oleoyl- and dioleoyl-MGlcDG, respectively [39]; an increase in the number of acyl chains with *cis* unsaturation thus increases the water-binding capacity. The maximum hydration also increases with increasing temperature and in the presence of divalent cations [44]. ^{2}H n.m.r. investigations of MGlcDG/^{2}H$_2$O mixtures have indicated that the water molecules bind to several sites on the lipid, probably by hydrogen bonding to the hydroxyl groups of the sugar ring [39]. The maximum hydration of dioleoyl-PC (DOPC) at 2 and 25 °C is 30–34 mol water per mol lipid [31, 46–48]. DOPE takes up about 18 mol water per mol lipid at 2 °C [48], while PE enriched in anteiso and iso-methyl-branched, saturated acyl chains takes up approximately 14 and 8 mol water per mol lipid, respectively, at 26 °C [49]. Thus, the maximum hydration of PE seems to be much more sensitive than that of MGlcDG to the acyl chain composition.

For comparison it can be mentioned that the maximum hydration at 35 °C for dielaidoyl-, palmitoyl/oleoyl- and dioleoyl-DGlcDG is approximately 7, 9, and 11 mol water, respectively per mol lipid [39]. Thus, although DGlcDG has two glucopyranosyl molecules as polar head-group, its ability to take up water is the same as that of MGlcDG. A plausible explanation of the behaviour of the glucolipids might be as follows. Note, first, that the non-ionic sugar lipids MGlcDG and DGlcDG, independent of the phase structure formed (H_{II} or L_α), have a much lower ability than e.g. the zwitterionic PC or the anionic PA to take up water [39]. The main reason for this is that, for the glucolipids, there are no strong repulsive forces caused by long-range columbic interactions. Furthermore, short-range repulsive forces are also much weaker for the non-ionic glucolipids than for PCs [39, 50]. Recently, Israelachvili and Wennerström [51] gave a possible explanation of this behaviour, in their discussion of the effect of steric forces between amphiphilic surfaces. Most probably, the steric or protrusion force between the glucolipid aggregates is diminished by the decreased mobility of the sugar head-groups as a result of hydrogen bonding (A. Holmgren, G. Lindblom & L. Rilfors, unpublished work). For PCs, on the other hand, the polar head-group moves rapidly and, consequently, the protrusion force will be larger and more water can be stored between the bilayers.

Monolayer studies at the air/water interface of the α- and β-anomers of dipalmitoyl-MGlcDG have shown that both the condensed and expanded areas of the α-linked anomers are greater than those of their β-linked counterparts. However, both anomers occupy smaller areas than dipalmitoyl-PC but larger areas than dipalmitoyl-PE [52]. From X-ray diffraction studies of an L_α phase formed by MGlcDG isolated from *A. laidlawii*, the area per molecule was calculated to be 6.1 and 6.3 nm^2 at 20 and 47 °C, respectively; the acyl chain composition of this MGlcDG was 30 molar % 16:0, 47 molar % 18:1c and 22 molar % short acyl chains ($n = 12$–15) [25].

The polar head-groups of the α- and β-anomers of ditetradecyl-MGlcDG have been studied by ^{2}H n.m.r. using specifically ^{2}H-labelled glucose [53–55]. Two assumptions had to be made in order to determine the molecular ordering of the head-groups with respect to the bilayer normal; first, the sugar ring is rigid and, second, there is a rapid rotation of this ring molecule about an axis to which an order parameter, S_{mol}, can be defined (this has to be done as no 'natural' axis of high symmetry exists in the sugar group). From the quadrupole splittings of three C—^{2}H bonds, which showed surprisingly similar values for the two anomers, S_{mol} was calculated to be 0.56 and 0.45 for the α- and β-anomers, respectively. From the ^{2}H n.m.r. data obtained Jarrell et al. [53–55] concluded that the orientation of the glucopyranose ring differs greatly for the two anomers. The sugar ring of the β-anomer is essentially fully extended away from the bilayer surface, while the pyranose ring of the α-anomer is almost parallel to the bilayer surface. In addition, from measurements of ^{2}H n.m.r. spin-relaxation times it was found that the rate of the head-group motion is slower in the α-anomer than in the β-anomer, which in turn has a slower head-group motion than that of PC and PG. Finally, it is interesting to note that the C_2—C_3 bond of the glycerol backbone of the β-anomer was estimated to be tilted away from the bilayer normal by about 3° [54]. Thus, the ordering of the glycerol backbone is consequently very similar to that reported for various phospholipids [56–59].

The order parameter profile (Fig. 2) and the transverse spin-relaxation rate have been determined for an L_α phase formed by MGlcDG containing a small fraction of biosynthetically incorporated perdeuterated 16:0 [25]. The order parameter profile is similar to those obtained from phospholipids [60]. Interestingly, the order of the acyl chains is higher in MGlcDG than in DGlcDG; DGlcDG, in contrast to MGlcDG, only forms L_β and L_α phases [39]. The transverse spin-relaxation data indicate the presence of slow reorientational motions, such as lipid lateral diffusion over a curved bilayer surface, the amplitude of which is directly related to the relative amount of MGlcDG in binary mixtures with DGlcDG [25]. Analogous results have been obtained from investigations of DOPC and DOPE and mixtures of these lipids; the order of the hydrocarbon chains is smaller in DOPC and is increased by the addition of DOPE [22, 23]. Consequently, two membrane lipids with the ability to form non-lamellar phases of the reversed type increase the order of the acyl chains when they are present in a bilayer.

It is interesting to note that, as the molecular ordering in a liquid crystalline phase depends on the 'shape' of the molecules in the lipid aggregate, the order parameter of the acyl chains may be taken as a measure of the packing of the lipid molecules. This has been illustrated recently (R. Thurmond, G. Lindblom & M. Brown, unpublished work) for some typical model membrane systems, where it was shown that the order profile is highest for an H_I phase and then decreases from the L_α phase to the H_{II} phase (Fig. 3).

Several authors have discussed the possibility of intermolecular hydrogen bonding between the polar head-groups of glycolipids [39, 41, 42, 61–63]. In crystals of glycolipids the molecules are packed so that intermolecular hydrogen bonds are formed [64, 65]. Unfortunately, it is difficult (see also above) to demonstrate conclusively the occurrence of hydrogen bonds in lipid/water systems [63]. However, possible intermolecular hydrogen bonding has been assumed in order to explain the following characteristic properties of glycolipids: the high packing density in monolayer films; the high temperatures (T_m) for the transition between L_β and L_α phases; the ability to form reversed non-lamellar phases; and the low water-binding capacity [61–63].

In summary, MGlcDG has similar, but not identical, physicochemical properties as PE: it has a larger area per molecule in a monolayer at the air/water interface; it has a lower maximum hydration; it is much more prone to form non-lamellar phases when it contains saturated acyl chains; it has slightly lower T_m values; and it has a larger enthalpy change associated with the L_β–L_α phase transition.

2-*O*-acyl-1-*O*-polyprenyl-α-D-glucopyranoside

This lipid is synthesized by *A. laidlawii* strain B and has an alcohol (probably a polyisoprenoid-based compound) glycosidically linked to the glucose moiety and a long-chain fatty acid esterified to the 2-hydroxyl group of glucose [66]. The phase behaviour of the lipid is markedly different in the heating and cooling modes and, in the former, depends also on the thermal history of the lipid/water sample [67]. Lipid dispersions that have been fully equilibrated at temperatures below 15 °C exhibit a conversion from a crystal-like lamellar gel phase to an H_{II} phase near 65 °C. However, upon cooling from high temperatures, the H_{II} phase persists to temperatures near 40 °C; a mixture of cubic and H_{II} phases seems to be present between 35 and 40 °C,

Fig. 2. Order parameter profiles at different ratios of MGlcDG: DGlcDG in the L_α phase

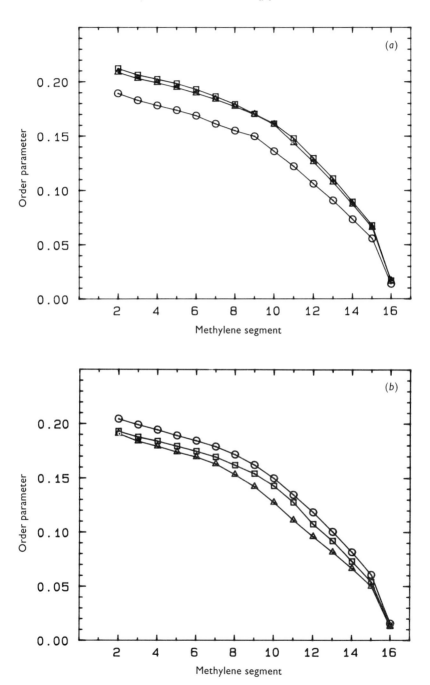

and an L_α phase is formed between 33 and 35 °C; below 33 °C the lipid dispersion transforms to a metastable lamellar gel phase. When this metastable gel phase is heated, an L_α phase forms at 33 °C and an H_{II} phase at 39 °C. The phase behaviour of the metastable gel phase is considered to reflect the physical properties of 2-O-acyl-1-O-polyprenyl-α-D-glucopyranoside when present in the *A. laidlawii* membrane [67]. Since the corresponding MGlcDG compound transforms from a lamellar to a cubic phase at 79 °C [42], the newly discovered glucolipid is more prone to form non-lamellar phases. A lipid with similar properties, 1,2-diacyl-3-O-[3-O-acyl-(α-D-glucopyranosyl)]-*sn*-glycerol has been found to occur also in *A. laidlawii* strain A (J. Hauksson, P.-O. Eriksson, G. Lindblom, L. Rilfors and Å. Wieslander, unpublished work).

Lipids forming non-lamellar phases affect the regulation of the lipid composition in microbial membranes

Stable non-bilayer structures cannot occur in a biological membrane, as this would destroy its barrier properties. We have found [13, 39] that, instead, the membrane lipids should be in the 'vicinity' of a non-lamellar phase transition under physiological conditions. Therefore the balance between bilayer-forming and non-bilayer-forming lipids is actively regulated in cells growing under different environmental conditions. This is supported by several experimental results [26, 39, 49, 68–70].

Acholeplasma laidlawii

The regulation of the lipid composition and the physicochemical properties of the membrane lipids have been investigated extensively for the plasma membrane of the parasite organism *A. laidlawii* ([13] and references therein). The membrane of this organism contains mainly six different lipids, three non-ionic and three ionic. MGlcDG and DGlcDG are the major lipids (55–75 molar %). The composition of the polar head-groups of the lipids is affected by several factors: the growth temperature; incorporation of different fatty acids into the lipids; and incorporation of sterols, hydrocarbons, alcohols, detergents, and chlorophyll into the membrane [39, 68, 69, 71, 72].

(a) The order parameter profile for DGlcDG containing perdeuterated 16:0 (DGlcDG-d_{31}), with varying amounts of MGlcDG added. ○, Pure DGlcDG-d_{31}; □, DGlcDG-d_{31}/MGlcDG (71:29, mol/mol); △, DGlcDG-d_{31}/MGlcDG (24:76, mol/mol). (b) The order parameter profile for MGlcDG containing perdeuterated 16:0 (MGlcDG-d_{31}), with varying amounts of DGlcDG added. ○, Pure MGlcDG-d_{31}; △, MGlcDG-d_{31}/DGlcDG (77:23, mol/mol); □, MGlcDG-d_{31}/DGlcDG (25:75, mol/mol). The water content was 9 mol ^1H_2O per mol lipid and the temperature was 48 °C. Reproduced from [25] with permission.

Fig. 3. Segmental order parameters, $S_{CD}^{(i)}$, as a function of the chain segment position, i

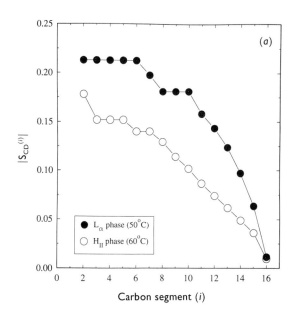

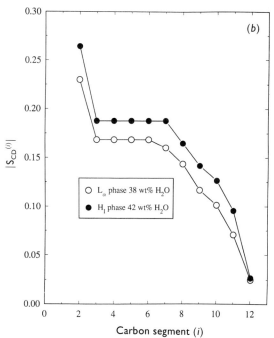

All of these factors have an influence on the molar ratio of MGlcDG:DGlcDG. As mentioned previously, MGlcDG has pronounced tendancies to form I and H_{II} phases, while DGlcDG and the other lipids form an L_α phase [39]. The following conclusion was drawn. When the growth conditions of the cells are changed in such a way that the phase equilibria of the membrane lipids are shifted towards non-lamellar structures, the cells respond by changing the ratio of MGlcDG:DGlcDG in order to stabilize the bilayer structure. Furthermore, it has been found that in most cases the formation of non-lamellar phases by *A. laidlawii* lipids depends critically upon the concentration of MGlcDG.

The regulation of the 'bilayer stability' is exemplified in Fig. 4. When the proportion of the oleoyl chains in the membrane lipids is increased, without changing the polar head-group composition, the phase equilibria are shifted towards the I and H_{II} phases [73, 74]. However, by lowering the molar ratio of MGlcDG:DGlcDG the transition to the non-lamellar phases begins at approximately the same temperature (10–15 °C above the growth temperature). An I phase appears in all the lipid extracts. Accordingly, two kinds of restrictions seem to determine the membrane lipid composition in the *A. laidlawii* cells. Firstly, a large proportion of the lipids are not allowed to enter the gel phase as this would inactivate the membrane-bound enzyme and transport activities. Secondly, non-lamellar lipid phases are not allowed to form permanently since this would break down the barrier properties of the membrane.

Clostridium butyricum

Similar experiments with *C. butyricum* have been performed [26]. The major membrane lipids in this organism are PE, plasmaenylethanolamine (PlaE), and the glycerol acetal of PlaE (GAPlaE). The ratio of (PE + PlaE):GAPlaE is decreased when the degree of acyl chain unsaturation is increased. The PE plus PlaE fraction forms an H_{II} phase above 0 °C, while GAPlaE forms an L_α phase up to at least 50 °C. Similar regulatory mechanisms thus seem to operate in *A. laidlawii* and *C. butyricum*, although the membrane lipids involved differ completely.

Bacillus megaterium

In *B. megaterium* the membrane lipids are PE, PG and DPG, containing mainly iso- and antesio-methyl-branched saturated acyl chains. With

---

Profiles are for (a) palmitoyl-d_{31}-linoleoyl-phosphatidylethanolamine in the L_α (●; 50 °C) and the H_{II} (○; 60 °C) phases; and (b) potassium laurate-d_{23} in the L_α (○; 38 wt% H_2O) and H_I (●; 42 wt% H_2O) phases. The order parameters were derived from ^2H n.m.r. quadrupole splittings. (R. Thurmond, G. Lindblom & M. F. Brown, unpublished work.)

increasing growth temperature the fraction of PE remains fairly constant (55–65 molar %), while the molar ratio of iso : anteiso acyl chains is markedly increased. PE isolated from *B. megaterium* grown at 20 °C has a low ratio of iso : anteiso acyl chains and begins to form an I phase at 50 °C at low water

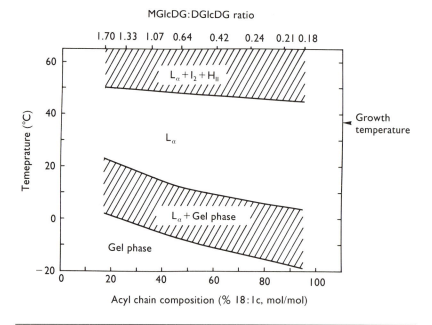

Fig. 4. Phase equilibria of total lipid mixtures, containing different amounts of 16:0 and 18:1c (lower x-axis), from membranes of *A. laidlawii* strain A-EF22 grown at 37 °C

Water contents were approximately 20 wt%. The upper x-axis shows the metabolically obtained MGlcDG:DGlcDG ratios. The lower hatched area denotes the gel–liquid crystalline phase transition interval as determined by e.s.r. The upper hatched area denotes the appearance of non-lamellar (H_{II} and/or cubic) phases in the lipid mixtures. Adapted from [39].

contents. PE isolated from cells grown at 55 °C has a 10-fold higher ratio of iso : anteiso acyl chains and forms an L_α phase up to at least 65 °C [49]. Since an increase in the temperature will increase the tendency of PE to form I and H_{II} phases, the temperature-induced regulation of the composition of the acyl chains in *B. megaterium* membranes is most probably necessary in order to maintain a stable lipid bilayer.

Escherichia coli

Recently, it was possible to apply our model to the regulation of the lipid composition in *E. coli*. Like some other organisms, *E. coli* coordinates the synthesis of membrane phospholipids to achieve a balanced composition that appears to be important for optimal growth. In wild-type cells under ordinary growth conditions, PE comprises about 70% of the total phospholipid, while PG makes up about 25% and DPG 5% [70, 75]. In *E. coli* the phospholipids are both the end-point in the metabolism, supplying the components of the membrane matrix, and precursors, supplying components for several metabolic pathways involved in macromolecular synthesis. Recently, it was found [70] that the gene encoding one of the enzymes responsible for the production of PE, the only zwitterionic phospholipid in the membrane of *E. coli*, could be inactivated, so that the membrane contained only anionic phospholipids, mainly PG and DPG. However, divalent metal ions such as Ca^{2+}, Mg^{2+} and Sr^{2+} were required in the growth medium in order for the bacterium to survive. It was also observed that there was no significant difference in the phospholipid:protein ratio in the membranes of the mutant strain relative to the wild-type strain. Previously, we have suggested [12] that the suppression of the growth phenotype of the mutant by Mg^{2+} (at the restrictive temperature) could be explained by the ability of DPG to substitute for PE, which forms an H_{II} phase. It should be noted that DPG in the presence of divalent ions also forms non-lamellar phases [76]. This explanation is analogous to the one given for the lipid regulation in *A. laidlawii*, based on the hypothesis that a lipid forming a non-lamellar phase structure is necessary for an optimal functioning membrane.

Pseudomonas fluorescens

The phase structures developed by PE isolated from *P. fluorescens* grown at 5 °C and 22 °C have been studied [77]. PE constitutes about 75% of the phospholipids in the membrane of this organism, the remainder being mainly PG and DPG. The proportion of unsaturated acyl chains in PE was decreased from 76% to 65% when the growth temperature was raised from 5 °C to 22 °C. PE (at 5 °C) produced a mixture of L_{α} and H_{II} phases from the growth temperature up to 43 °C, while above this temperature only an H_{II} phase was present. PE (at 22 °C) formed an H_{II} phase at 65 °C and a mixed-phase system between the growth temperature and 65 °C. Surface pressure–area curves for monolayers of the two PE preparations remained unchanged down to 5 °C, indicating that T_m is below 5 °C for both the preparations [77]. It can be concluded that the increased tendency of PE to form non-lamellar phases at higher temperatures is counteracted by regulating the acyl chain composition, and the balance between the lamellar and non-lamellar phases is thus kept approximately constant [77]. Recently, it was also reported that a lipid extract from membranes of this organism constitutes non-lamellar phases, and a pseudophase diagram was constructed by considering the lipid extract as one component [78].

Non-lamellar phases formed by some anionic membrane lipids

Anionic lipids are present in all biological membranes, PG and DPG being the predominant anionic lipids of bacterial membranes. These two lipids together with PS are common in eukaryotic cell membranes. PA is a minor component of cell membranes but is a key intermediate in the biosynthesis of other phosphoglycerides. Certainly, a detailed knowledge of the physicochemical properties of anionic membrane lipids, like head-group hydration, counterion association, lipid aggregate structure, monolayer curvature and the phase behaviour, will give us a better understanding of biological processes in which anionic lipids are involved. So far comparatively little has been reported on the ionic lipids.

Because of the net negative charge of anionic lipids at physiological pH values, the interaction with monovalent and divalent cations, as well as changes of the pH value in these lipid/water systems, will affect the phase equilibria. The unique behaviour of many charged lipids to swell and form unilamellar vesicles in excess water has been reviewed by Hauser [79]. The T_m value is profoundly increased for PG and PS by addition of Ca^{2+} [80, 81]. This transition temperature is also increased for PG and PS for a change from neutral to acidic pH values [80–84]. When the pH value is decreased, a monolayer of PS molecules will condense [85] and the orientational order of the acyl chains in a PS bilayer is increased [86]. A transition from an L_α to an H_{II} phase is induced by the addition of Ca^{2+}, Mg^{2+} or Mn^{2+} to DPG and PA [87–90], by the addition of Na^+ to DPG [91] and by the addition of Li^+ to PS [92]. The L_α–H_{II} phase transition also occurs for PS, PA and DPG when the pH is decreased from a neutral to an acidic value [86, 90, 91, 93–95]. Changes in the concentration of cations and the pH can induce fusion of anionic lipid vesicles [96] and affect the lipid composition in biological membranes [97, 98].

Recently, we reported [98] a study of the phase equilibria and the counterion association of the sodium salts of DOPA, DOPS, DOPG and DPG, in which we used X-ray diffraction, and ^2H, ^{23}Na and ^{31}P n.m.r. at water concentrations ranging from 10–98 wt% of ^2H_2O. DOPA behaves differently to the other three lipids in forming an H_{II} phase up to about 35 wt% of water (Fig. 5). Above approximately 50 wt% of water the counterion concentration close to the lipid–water interface is constant and very high in all the anionic lipid/water systems studied. This behaviour was explained in terms of ion condensation [98].

Phase equilibria

A qualitative understanding of the phase behaviour of lipid/water systems can be arrived at using simple, but most useful, theoretical models. The phase structures may be controlled by the spontaneous radius of curvature of the lipid monolayer. Lipid compositions that can adopt negative values of this curvature (i.e. lipid monolayers curved around water regions) form stable, reversed liquid crystalline phases [16]. The ability of a lipid monolayer to adopt negative radii of curvature is determined in part by the ability of the head-groups to be packed closely together: the area per lipid molecule is

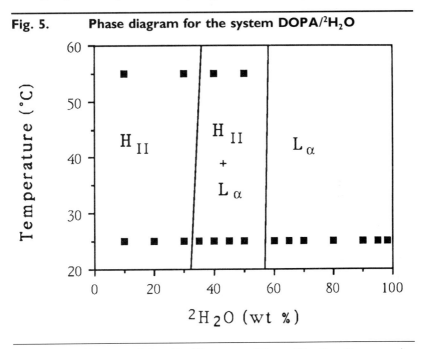

Fig. 5. Phase diagram for the system DOPA/^{2}H$_2$O

The phase equilibria were deduced by ^{31}P n.m.r. and X-ray diffraction. The reversed hexagonal and the lamellar liquid crystalline phases are denoted by H_{II} and L_α, respectively. Reproduced from [98] with permission.

smaller at the head-group than in the hydrocarbon chain region for a monolayer with negative curvature. One would think that a lipid with a charged head-group would be less prone to adopt such a closely packed configuration as a result of electrostatic repulsion between the head-groups. Hence, these lipids should have less of a tendency to adopt reversed-phase structures than uncharged lipids with otherwise similar chemical structures. Steric interactions between head-groups must also affect their ability to pack closely together. Therefore, we might expect lipids with bulkier head-groups to be less prone to form reversed phases.

To account for these tendencies, Israelachvili and coworkers [99] defined a packing parameter in terms of the hydrocarbon chain volume, the area per molecule at the lipid–water interface and the length of the hydrocarbon chains. However, the value of each of these quantities depends upon the nature of the phase in which the lipid is found, and upon the temperature and ambient conditions. For instance, the value of the packing parameter for DOPE changes significantly over a small temperature range [40, 100]. Therefore, with this formalism it is difficult to make explicit predictions about the temperature- and composition-dependence of lipid-phase behaviour. For explicit calculations of the relative stability of lamellar and reversed phases, a phenomenological model of material physics introduced by Helfrich [101] has great advantages, although the molecular details are concealed. In this model, the lipid aggregate stability is dominated by an elastic free energy of curvature, and a concept related to the packing parameter, namely the so-called spontaneous curvature of the lipid monolayer, is utilized [101, 102].

Gruner and coworkers have shown that the energy of curvature plays a dominant role in the formation of H_{II} phases for some zwitterionic phospholipids. For a phospholipid monolayer to form a cylinder there will also be a non-zero packing energy [102]. In particular, the hydrocarbon chains of the phospholipid molecules must stretch to fill the interstitial regions between the cylindrical aggregates building up the H_{II} phase. The smaller the radius of the water cylinder, the smaller the hydrophobic interstices and the easier it is for the lipid acyl chains to elongate to fill the volume [14, 16, 102, 103].

Dioleoyl-phosphatidic acid
As can be inferred from Fig. 5 an H_{II} phase is formed for the DOPA/water system up to a water content of about 35 wt%, and an L_α phase can swell up to a water concentration of at least 95 wt%. When a large fraction of DOPA molecules is neutralized by association of counterions, the head-groups can be packed quite closely together. Therefore, the lipid monolayer is able to adopt a negative curvature, leading to the formation of an H_{II} phase. With increasing water concentration in the DOPA/water system, the interstitial regions between the growing lipid/water cylinders, building up the H_{II} phase, will increase. The creation of such void volumes is energetically unfavourable, and will eventually lead to the induction of an L_α phase at a certain water content (Fig. 5). When the water concentration is increased further the L_α phase swells, reaching large distances, as a result of the long-range coulombic forces, between the charged bilayers. It is well known that even a small increase in the surface charge density may drastically increase the tendency for water to be taken up by a lamellar phase [104, 105]. The thickness of the water layer at 98 wt% of water for the DOPA system can be estimated to be approximately 200 nm.

Diphosphatidylglycerol
DPG forms H_{II} and I phases at low water contents and temperatures above 45–50 °C [87, 98]. Cubic phases are often located between L_α and H_{II} phases in lipid/water phase diagrams. One can understand the formation of the cubic phase structure by considering the lipid monolayer curvature as being balanced by the tendency of the molecules to adopt a wedge-like shape and

the packing constraints of the hydrocarbon chains in the bilayer [13]. DPG consequently has a greater tendency to form non-lamellar phases than DOPS and DOPG, which form just an L_α phase between 10 and 98 wt% water and up to at least 55 °C. This conclusion is reasonable because DOPG and DOPS have bulkier head-groups than DPG. The acyl chain composition of DPG affects the phase equilibria in an expected way; bovine heart DPG with 89 molar% linoleoyl chains forms an H_{II} phase above 0 °C, while DPG isolated from *Bacillus subtilis*, containing 99 molar% straight or methyl-branched saturated acyl chains, transforms to an H_{II} phase between 40 °C and 50 °C [106].

PS has been shown to form an H_{II} phase both in the anhydrous state [107, 108] and, at low pH values and in the presence of Li^+, at high water contents. PG seems to have no tendencies to form non-lamellar phases; it forms an L_α phase even in the presence of divalent cations [109]. The ability of anionic lipids to form non-lamellar phases can be graded as follows [98]: PA > DPG > PS > PG. The phase behaviour of DOPA and DPG shows that it cannot be assumed *per se* that phospholipids having charged head-groups only form an L_α phase together with pure water.

It is well documented that the addition of anionic phospholipids to an H_{II} phase formed by PE or MGlcDG breaks up this phase and converts it to an L_α phase [12]. However, addition of DOPG to an H_{II} phase formed by DOPC-*n*-dodecane/water did not lead to a phase transition [98]; an H_{II} phase was obtained even by a DOPG-*n*-dodecane/water mixture. It has been shown previously that the packing constraints of the hydrocarbon chains, which oppose the formation of non-lamellar phases, are removed by the presence of the alkane [29, 31, 102], and even a lipid such as PG then forms an H_{II} phase.

Recent spectroscopic studies of some cubic phases

An extensive review of cubic phases formed by membrane lipids and their possible biological relevance was published by our group in 1989 [13]. For a more detailed discussion of such cubic phases the reader is referred to that survey and here we will only briefly report on the most recent results we have obtained on the structure and dynamics of cubic phases.

Cubic phases formed at high water contents
Lipids in water form a wide variety of liquid crystalline phases, of which the cubic phase structures are the most complex [13, 15, 16, 110–112]. In recent years a relatively large number of investigations of the cubic phases located between the aqueous micellar solution and the H_I phase have been reported [110, 113–118] (E. Berggren, P.-O. Westlund, P.-O. Eriksson & G. Lindblom, unpublished work). The first observation of a cubic phase with this location was made by Reiss-Husson [119] for egg lysolecithin in water.

Lysolecithins with hydrocarbon chain lengths between 12 and 16 carbons have also been shown to form this type of cubic phase [114, 116, 120], and X-ray diffraction has shown the space group to be $Pm3n/P43n$ [113, 118, 121, 122]. However, the structure (Fig. 1d) of these cubic phases has been debated [110, 113, 114, 116, 123].

The main dispute has been whether this cubic phase is bicontinuous or consists of closed micellar aggregates [110, 114]. We are faced with a problem that cannot be solved by X-ray diffraction alone, but requires other methods [13]. Obviously, such methods should be sensitive to the restriction of lipid molecules to be localized in a limited space, or aggregate, in the phase structure under study. The necessary prerequisites for such investigations are indeed provided by the n.m.r. diffusion technique, which we have used frequently. The main advantage of this method is that the translational diffusion coefficient of the lipid or water can be measured directly, i.e. no probe molecule has to be involved and no model-dependent assumptions have to be made. The time during which the translational motion is observed can be varied over a wide scale corresponding to motional distances of macroscopical orders of magnitude (μm). Thus, molecular transport over distances much larger than the dimension of a single micelle can be observed.

From measurements of lipid diffusion in various liquid crystalline phases it is thus possible to distinguish between cubic phases built up of closed aggregates and continuous hydrocarbon regions ([13] and references therein). The simple idea behind the method is thus based on the fact that, for a cubic phase consisting of micelles, the lipid molecules will be restricted to diffusing within the micelle and can therefore only move a very short distance, whereas in a bicontinuous phase it can perform translational diffusion over large distances. The same basic idea can also be used to interpret diffusion coefficients determined by a fluorescent probe method such as fluorescence recovery after photobleaching (FRAP) ([13]; see also [124]) to discriminate between the fundamentally different cubic phase structures. It should be noted, however, that a probe molecule may perturb the aggregate size and shape (G. Orädd, G. Lindblom, L. B.-Å. Johansson & G. Wikander, unpublished work); in particular this might occur with the usually very large fluorescent molecules used in the FRAP experiments. Being aware of this problem, another probe method which utilizes fluorescence quenching has great advantages in studies of cubic phases. An aggregation number of the possibly present micelles may be obtained and, in most cases, information may be extracted about the occurrence of large hydrocarbon regions in the cubic structure. Recently, we carried out some studies with time-resolved fluorescence quenching to obtain further support for the previously suggested cubic phase structure of lysolecithins composed of slightly distorted globular micelles (G. Lindblom, L. B.-Å. Johansson, P.-O. Eriksson, G. Arvidson & G. Wikander, unpublished work).

In most fluorescence-quenching studies of amphiphilic systems the aggregation numbers, N, of micelles have been determined. Recently, it was shown that the aggregate size of cubic liquid crystals may also be determined [117] (G. Lindblom *et al.*, unpublished work). The aggregation numbers for the micellar aqueous solution and the cubic I_1 phase of the lauroyllysophosphatidylcholine (LaLPC)/water system at 25–30 °C were found to be $N_{\text{micelle}} = 77 \pm 15$ and $N_{\text{cubic}} = 89 \pm 20$.

When the observed aggregation number of a micelle becomes larger than a value possible for a spherical aggregate, the actual micelles, of course, must have a different shape, which is limited in at least one dimension because of the length of the hydrocarbon chain. Thus, cylindrical and disc-like structures are the only conceivable forms. With increasing concentrations of LaLPC, the I_1 phase is followed by an H_1 phase, indicating a tendency for a rod-like micellar structure to be formed in the I_1 phase. The aggregation number for the micellar solution is comparable with the value obtained for the I_1 phase, strongly indicating that the cubic structure consists of short rod-like micelles [114, 116, 117].

Besides the very important n.m.r. diffusion results, showing conclusively that the cubic phase consists of closed micelles, are the extraordinary line shapes obtained for ^{2}H, ^{14}N and, in particular, for the ^{31}P n.m.r. spectra of this cubic phase [114, 116]. It was suggested [114] that these n.m.r. line shapes are caused by the rotational motion of anisometric micelles in the cubic unit cell, so that two of the micelles (in the corner and in the middle of the unit cell) perform an isotropic motion, while the other six micelles (on the surfaces of the unit cell) are restricted to anisotropic rotational motion (see Fig. 1*d*). This interpretation was based on the assumption that the micelles have a non-spherical shape, which has now been confirmed by the present fluorescence quenching data described above.

Recently, a quantitative line shape analysis of the ^{2}H, ^{14}N and ^{31}P n.m.r. spectra of lysolecithin in the cubic phase was performed (E. Berggren *et al.*, unpublished work). The line shape calculation was based on the dynamic model suggested previously [114], using a stochastic Liouville approach. The slowest dynamical motion of the micelles was found to be in the slow-motion regime, and this was taken into account by a Brownian rotational diffusion model. The line shape calculations confirm that the cubic I_1 phase consists of two dynamically different types of aggregates. The chemical shift anisotropy in the ^{31}P n.m.r. spectrum is of crucial importance, indicating that the six anisometric micelles, located on the surfaces of the unit cell, are performing a rotational diffusion restricted to occurring within that plane. As mentioned above, the remaining two anisometric micelles in the corners and in the middle of the unit cell, are subjected to isotropic rotational motion.

Charvolin and Sadoc [123] suggested that every lyotropic structure can be described in the same geometrical terms, as periodically ordered systems of fluid films separated by interfaces. The structures of these films must reconcile constant interfacial distances and curvatures. When the conditions are such that the interfaces become curved, a typical geometrical frustration arises, a problem which is solved by introducing structure of disclinations [123]. In their model for a cubic phase, formed for instance by lysolecithins at high water content, an assembly of two dodecahedra and six tetrakaidecahedra is obtained in which the micelles reside. There is a striking similarity between the structure model we use and the one proposed by Charvolin and Sadoc. Finally, it is interesting to note that this cubic phase containing palmitoyllysolecithin was recently used [125] for u.v./visible and circular dichroism investigations of an immobilized enzyme (α-chymotrypsin). The enzyme was incorporated in the cubic phase structure without affecting the physical properties of the liquid crystalline phase.

N.m.r.-diffusion studies of molecules solubilized in a bicontinuous cubic phase

In many respects, the phase behaviour presented by mono-olein (MO) itself, and in mixtures with membrane lipids, provides us with excellent model membrane systems for physicochemical investigations of structure, dynamics and lipid–lipid interactions. The most striking property observed in the binary phase diagram of MO and water is that a bicontinuous cubic phase is in equilibrium with water [126, 127]. The lipid diffusion coefficient in the cubic phase is close to two thirds of the lipid diffusion in the lamellar phase [127], a relationship which is to be expected for a cubic phase built from bilayer units [127, 128]. Therefore, a structure based on lamellar units was proposed [127]. The cubic phases of monoglycerides and water can solubilize a substantial amount of hydrophobic and amphiphilic molecules [46] (P.-O. Eriksson & G. Lindblom, unpublished work). From the observation of a discontinuity in the measured water diffusion coefficient in the cubic phase region of the MO/water phase diagram, it was concluded that two cubic phases with different structures are present. It was also found that an increasing degree of acyl chain unsaturation gives a significant increase in the lipid diffusion coefficient. In this work it was also possible for the first time to simultaneously determine the lipid diffusion coefficients of two lipid molecules residing in the same cubic phase structure. In particular it was observed that the diffusion coefficient of MO is approximately a factor of 2 larger than the diffusion coefficient of DOPC, and that both diffusion coefficients decrease with increasing DOPC content. Similarly, addition of cholesterol, gramicidin and oleoyllysophosphatidylcholine retarded the MO diffusional motion in the cubic phase, whereas the addition of monostearoylglycerol had no effect on the MO diffusion. It was concluded that the translational diffusion coefficients are mutually affected by the motion of the molecules building up the lipid aggregates. It is interesting to note that in previous n.m.r. diffusion measurements on macroscopically aligned lamellar PC/water phases, it was found that the addition of cholesterol had a very small effect on the diffusion coefficient of the different PCs studied [129]. If anything, a slight increase in the PC translational diffusion was observed with increasing cholesterol content in the bilayers. This most probably means that the cholesterol diffusion coefficient is very similar to or slightly larger than that of PC.

Fourier transform i.r. studies of cubic phases containing MO

Major advances in the technique of fourier transform i.r. spectroscopy over the last few years have resulted in a considerable number of publications of the application of this method to membrane lipid systems. It is a very powerful non-perturbing technique with which all parts of a molecule can be studied. Thus, for example, the hydrocarbon chain conformation of phospholipids can be determined quantitatively [130], and information about phospholipid phase transitions [131] and hydrogen bonding in different liquid crystalline phases can be obtained [132-134]. From a Fourier-transform i.r. linear dichroism investigation of a macroscopically aligned lamellar liquid crystalline sample, the molecular ordering of different groups of the molecules can be obtained through the order parameter [132]. Hitherto, our Fourier transform i.r. studies have embraced lipid systems containing monoglycerides. For example, the method has been used to delineate the importance of the hydrogen bonding between lipid head-groups in a system containing DOPC and MO [134], where the polymorphism is rather rich [46]. It has been shown that the polar head-group of MO is intramolecularly hydrogen bonded to the carbonyl group [133]. Both MO and DOPC form an L_α phase with a water content of about 17 wt%. Addition of DOPC to an L_α phase of the MO/water system results in the intramolecular hydrogen bond of MO being broken and an intermolecular hydrogen bond with the polar head-group of DOPC being formed. This hydrogen bonding contributes to a decrease in the mean interfacial surface area per lipid molecule, while the acyl chain volume can be considered as being constant. Thus, the effective shape of the molecules will change towards a more wedge-like geometry if MO or DOPC is added to the L_α phase. This will eventually lead to a phase transition to a non-lamellar structure. In terms of monolayer bending elasticity, this means that the lipid monolayer adopts an increasingly negative curvature as the amount of MO or DOPC increases in the MO/DOPC/water system. The experimental finding that the carbonyl group is less hydrated in the cubic phase than in the lamellar phase strongly supports this interpretation. Water penetrating into the lipid monolayer close to the acyl region will therefore lead to an increase in the interfacial tension, which will be counteracted by a change to a more negative curvature of the monolayer. Thus, the system responds to this unfavourable increase in free energy by curling the lipid monolayer to prevent hydrocarbon-water contact. Furthermore, attractive interactions between the polar head-groups will lead to a change in the molecular shape to a more wedge-like geometry, also imposing a curling of the lipid monolayer. It was concluded [134] that hydrogen bonding plays an important role in modifying the spontaneous curvature and the molecular packing of the lipid phase structures.

This work was supported by the Swedish Natural Science Research Council.

References

1. Singer, S. J. & Nicolson, G. L. (1972) Science 175, 720-730
2. Chapman, D. (1958) J. Chem. Soc. 152, 784-789
3. Chapman, D. & Salsbury, N. K. (1966) Trans. Farad. Soc. 62, 2607-2621
4. Chapman, D., Byrne, P. & Shipley, G. G. (1966) Proc. R. Soc. London A 290, 115-142
5. Chapman, D., Williams, R. M. & Ladbrooke, B. D. (1967) Chem. Phys. Lipids 1, 445-475
6. Chapman, D., Owen, N. F., Phillips, M. C. & Walker, D. A. (1969) Biochim. Biophys. Acta 183, 458-465
7. Ladbrooke, B. D. & Chapman, D. (1989) Chem. Phys. Lipids 3, 304-367
8. Oldfield, E. & Chapman, D. (1972) FEBS Lett. 23, 285-297
9. Oldfield, E., Keough, K. & Chapman, D. (1972) FEBS Lett. 20, 344-346
10. Veksli, Z., Salsbury, N. J. & Chapman, D. (1969) Biochim. Biophys. Acta 183, 434-445
11. Chapman, D. & Oldfield, E. (1974) Methods Enzymol. 32, 198-211
12. Rilfors, L., Lindblom, G., Wieslander, Å. & Christiansson, A. (1984) Biomembranes 12, 205-245
13. Lindblom, G. & Rilfors, L. (1989) Biochim. Biophys. Acta 988, 221-256
14. Lindblom, G. & Rilfors, L. (1990) in NATO ASI Series, Vol. H 40; Dynamics and Biogenesis of Membranes (Op den Kamp, J. A. F., ed.), pp. 43-64, Springer Verlag, Berlin
15. Seddon, J. M. (1990) Biochim. Biophys. Acta 1031, 1-69
16. Tate, M. W., Eikenberry, E. F., Turner, D. C., Shyamsunder, E. & Gruner, S. M. (1991) Chem. Phys. Lipids 57, 147-164
17. Taraschi, T. F., De Kruijff, B., Verkleij, A. & Echteld, C. J. A. (1982) Biochim. Biophys. Acta 685, 153-161
18. Siegel, D. P. (1986) Chem. Phys. Lipids 42, 279-301
19. Blume, A., Wittebort, R. J., DasGupta, S. K. & Griffin, R. G. (1982) Biochemistry 21, 6243-6253
20. Marsh, D., Watts, A. & Smith, I. C. P. (1983) Biochemistry 22, 3023-3026
21. Perly, B., Smith, I. C. P. & Jarrell, H. C. (1985) Biochemistry 24, 4659-4665
22. Cullis, P. R., Hope, M. J. & Tilcock, C. P. S. (1986) Chem. Phys. Lipids 40, 127-144
23. Lafleur, M., Bloom, M. & Cullis, P. R. (1990) Biochem. Cell Biol. 68, 1-8
24. Thurmond, R. L., Lindblom, G. & Brown, M. F. (1990) Biochem. Biophys. Res. Commun. 173, 1231-1238
25. Eriksson, P.-O., Rilfors, L., Wieslander, Å., Lundberg, A. & Lindblom, G. (1991) Biochemistry 30, 4916-4924
26. Goldfine, H., Johnston, N. C., Mattai, J. & Shipley, G. G. (1987) Biochemistry 26, 2814-2822
27. DeChavigny, A., Heacock, P. N. & Dowhan, W. (1991) J. Biol. Chem. 266, 5323-5332
28. Jensen, J. W. & Schutzbach, J. S. (1989) Biochemistry 28, 851-855
29. Sjölund, M., Rilfors, L. & Lindblom, G. (1989) Biochemistry 28, 1323-1329
30. Abrahamsson, S., Pascher, I., Larsson, K. & Karlsson, K.-E. (1972) Chem. Phys. Lipids 8, 152-179
31. Sjölund, M., Lindblom, G., Rilfors, L. & Arvidson, G. (1987) Biophys. J. 52, 145-153
32. Bryszewska, M. & Epand, R. M. (1988) Biochim. Biophys. Acta 943, 485-492
33. Epand, R. M. & Bottega, R. (1987) Biochemistry 26, 1820-1825
34. Thurmond, R. L., Lindblom, G. & Brown, M. F. (1991) Biophys. J. 60, 728-732
35. Ulmius, J., Lindblom, G., Wennerström, H., Johansson, L. B.-Å., Fontell, K., Söderman, O. & Arvidson, G. (1982) Biochemistry 21, 1553-1560
36. Nichols, J. W. & Ozarowski, J. (1990) Biochemistry 29, 4600-4606
37. Batenburg, A. M., Bougis, P. E., Rochat, H., Verkleij, A. J. & De Kruijff, B. (1985) Biochemistry 24, 7101-7110
38. Batenburg, A. M., Hibbeln, J. C. L., Verkleij, A. J. & De Kruijff, B. (1987) Biochim. Biophys. Acta 903, 142-154
39. Lindblom, G., Brentel, I., Sjölund, M., Wikander, G. & Wieslander, Å. (1986) Biochemistry 25, 7502-7510
40. Tilcock, C. P. S. & Cullis, P. R. (1982) Biochim. Biophys. Acta 684, 212-218
41. Mannock, D. A., Lewis, R. N. A. H., Sen, A. & McElhaney, R. N. (1988) Biochemistry 27, 6852-6859
42. Mannock, D. A., Lewis, R. N. A. H. & McElhaney, R. N. (1990) Biochemistry 29, 7790-7799
43. Sen, A., Hui, S.-W., Mannock, D. A., Lewis, R. N. A. H. & McElhaney, R. N. (1990) Biochemistry 29, 7799-7804

44. Wieslander, Å., Ulmius, J., Lindblom, G. & Fontell, K. (1978) Biochim. Biophys. Acta 512, 241–253
45. Huang, F., Zhao, B.-Z., Wen, D.-C. & Qian, Y.-M. (1989) Chinese Sci. Bull. 34, 514–517
46. Gutman, H., Arvidson, G., Fontell, K. & Lindblom, G. (1984) in Surfactants in Solution (Mittal, K. & Lindman, B., eds.), Vol. 1, pp. 143–152, Plenum Press, New York
47. Bergenståhl, B. A. & Stenius, P. (1987) J. Phys. Chem. 91, 5944–5948
48. Gruner, S. M., Tate, M. W., Kirk, G. L., So, P. T. C., Turner, D. C., Keane, D. T., Tilcock, C. P. S. & Cullis, P. R. (1988) Biochemistry 27, 2853–2866
49. Rilfors, L., Khan, A., Brentel, I., Wieslander, Å. & Lindblom, G. (1982) FEBS Lett 149, 293–298
50. Marra, J. (1986) J. Colloid Interface Sci. 109, 11–20
51. Israelachvili, J. N. & Wennerström, H. (1990) Langmuir 6, 873–876
52. Asgharian, B., Cadenhead, D. A., Mannock, D. A., Lewis, R. N. A. H. & McElhaney, R. N. (1989) Biochemistry 28, 7102–7106
53. Jarrell, H. C., Giziewicz, J. B. & Smith, I. C. P. (1986) Biochemistry 25, 3950–3957
54. Jarrell, H. C., Jovall, P. Å. Giziewicz, J. B., Turner, L. A. & Smith, I. C. P. (1987) Biochemistry 26, 1805–1811
55. Jarrell, H. C., Wand, A. J., Giziewicz, J. B. & Smith, I. C. P. (1987) Biochim. Biophys. Acta 897, 69–82
56. Wohlgemuth, R., Waespe-Sarcevic, N. & Seelig, J. (1980) Biochemistry 19, 3315–3321
57. Browning, J. L. & Seelig, J. (1980) Biochemistry 19, 1262–1270
58. Gally, H.-U., Pluschke, G., Overath, P. & Seelig, J. (1981) Biochemistry 20, 1825–1831
59. Strenk, L. M., Westerman, P. W. & Doane, J. W. (1985) Biophys. J. 48, 765–773
60. Seelig, J. & Seelig, A. (1980) Q. Rev. Biophys. 13, 19–61
61. Hinz, H.-J., Six, L., Ruess, K.-P. & Liefländer, M. (1985) Biochemistry 24, 806–813
62. Curatolo, W. (1987) Biochim. Biophys. Acta 906, 111–136
63. Boggs, J. M. (1987) Biochim. Biophys. Acta 906, 353–404
64. Moews, P. C. & Knox, J. R. (1976) J. Am. Chem. Soc. 98, 6628–6633
65. Pascher, I. & Sundell, S. (1977) Chem. Phys. Lipids 20, 175–191
66. Bhakoo, M., Lewis, R. N. A. H. & McElhaney, R. N. (1987) Biochim. Biophys. Acta 922, 34–45
67. Lewis, R. N. A. H., Yue, A. W. B., McElhaney, R. N., Turner, D. C. & Gruner, S. M. (1990) Biochim. Biophys. Acta 1026, 21–28
68. Wieslander, Å., Christiansson, A., Rilfors, L. & Lindblom, G. (1980) Biochemistry 19, 3650–3655
69. Wieslander, Å., Rilfors, L. & Lindblom, G. (1986) Biochemistry 25, 7511–7517
70. DeChavigny, A., Heacock, P. N. & Dowhan, W. (1991) J. Biol. Chem. 266, 5323–5332
71. Rilfors, L., Wikander, G. & Wieslander, Å. (1987) J. Bacteriol. 169, 830–838
72. Wieslander, Å & Selstam, E. (1987) Biochim. Biophys. Acta 901, 250–254
73. Wieslander, Å., Rilfors, L., Johansson, L. B.-Å. & Lindblom, G. (1981) Biochemistry 20, 730–735
74. Khan, A., Rilfors, L., Wieslander, Å. & Lindblom, G. (1981) Eur. J. Biochem. 116, 215–220
75. Raetz, C. R. H. & Dowhan, W. (1990) J. Biol. Chem. 265, 1235–1238
76. Rilfors, L., Eriksson, P.-O., Arvidson, G. & Lindblom, G. (1986) Biochemistry 25, 7702–7711
77. Cullen, J., Phillips, M. C. & Shipley, G. G. (1971) Biochem. J. 125, 733–742
78. Mariani, P., Rivas, E., Luzzati, V. & Delacroix, H. (1990) Biochemistry 29, 6759–6810
79. Hauser, H. (1984) Biochim. Biophys. Acta 772, 37–50
80. Verkleij, A. J., De Kruijff, B., Ververgaert, P. H. J. T., Tocanne, J. F. & Van Deenen, L. L. M. (1974) Biochim. Biophys. Acta 339, 432–437
81. Jacobson, K. & Papahadjopoulos, D. (1975) Biochemistry 14, 152–161
82. Mombers, C., Van Dijck, P. W. M., Van Deenen, L. L. M., De Gier, J. & Verkleij, A. J. (1977) Biochim. Biophys. Acta 470, 152–160
83. Findlay, E. J. & Barton, P. G. (1978) Biochemistry 17, 2400–2405
84. Van Dijck, P. W. M., De Kruijff, B., Verkleij, A. J., Van Deenen, L. L. M. & De Gier, J. (1978) Biochim. Biophys. Acta 512, 84–96
85. Demel, R. A., Paltauf, F. & Hauser, H. (1987) Biochemistry 26, 8659–8665
86. De Kroon, A. I. P. M., Timmermans, J. W., Killian, J. A. & De Kruijff, B. (1990) Chem. Phys. Lipids 54, 33–42
87. Rand, R. P. & Sengupta, S. (1972) Biochim. Biophys. Acta 255, 484–492
88. Cullis, P. R., Verkleij, A. J. & Ververgaert, P. H. J. T. (1978) Biochim. Biophys. Acta 513, 11–20
89. Van Venetië, R. &. Verkleij, A. J. (1981) Biochim. Biophys. Acta 645, 262–269
90. Verkleij, A. J., De Maagd, R., Leunissen-Bijvelt, J. & De Kruijff, B. (1982) Biochim. Biophys. Acta 684, 255–262
91. Seddon, J. M., Kaye, R. D. & Marsh, D. (1983) Biochim. Biophys. Acta 734, 347–352

92. Cevc, G., Seddon, J. M. & Marsh, D. (1985) Biochim. Biophys. Acta **814**, 141-150
93. Hope, M. J. & Cullis, P. R. (1980) Biochem. Biophys. Res. Commun. **92**, 846-852
94. Harlos, K. & Eibl, H. (1981) Biochemistry **20**, 2888-2892
95. Farren, S. B. & Cullis, P. R. (1980) Biochem. Biophys. Res. Commun. **97**, 182-191
96. Papahadjopoulos, D., Vail, W. J., Newton, C., Nir, S., Jacobson, K., Poste, G. & Lazo, R. (1977) Biochim. Biophys. Acta **465**, 579-598
97. Op den Kamp, J. A. F., Redai, I. & Van Deenen, L. L. M. (1969) J. Bacteriol. **99**, 298-303
98. Lindblom, G., Rilfors, L., Hauksson, J., Brentel, I., Sjölund, M. & Bergenståhl, B. (1991) Biochemistry **30**, 10938-10948
99. Israelachvili, J. N., Marcelja, S. & Horn, R. G. (1980) Q. Rev. Biophys. **13**, 121-200
100. Kirk, G. L. & Gruner, S. M. (1985) J. Physique **46**, 761-769
101. Helfrich, W. (1973) Z. Naturforsch. Teil C**28**, 693-703
102. Gruner, S. M. (1985) Proc. Natl. Acad. Sci. U.S.A. **82**, 3665-3669
103. Tate, M. W. & Gruner, S. M. (1987) Biochemistry **26**, 231-236
104. Gulik-Krzywicki, T., Tardieu, A. & Luzzati, V. (1969) Mol. Cryst. Liq. Cryst. **8**, 285-291
105. Brentel, I., Arvidson, G. & Lindblom, G. (1987) Biochim. Biophys. Acta **904**, 401-404
106. Vasilenko, I., De Kruijff, B. & Verkleij, A. (1982) Biochim. Biophys. Acta **684**, 282-286
107. Williams, R. M. & Chapman, D. (1970) Prog. Chem. Fats Lipids **11**, 1-79
108. Hauser, H., Paltauf, F. & Shipley, G. G. (1982) Biochemistry **21**, 1061-1067
109. Farren, S. B. & Cullis, P. R. (1980) Biochem. Biophys. Res. Commun. **97**, 182-191
110. Mariani, P., Luzzati, V. & Delacroix, H. (1988) J. Mol. Biol. **204**, 165-189
111. Larsson, K. (1989) J. Phys. Chem. **93**, 7304-7314
112. Fontell, K. (1990) Colloid Polym. Sci. **268**, 264-285
113. Fontell, K., Fox, K. & Hansson, E. (1985) Mol. Cryst. Liq. Cryst. **1**, 9-17
114. Eriksson, P.-O., Lindblom, G. & Arvidson, G. (1985) J. Phys. Chem. **89**, 1050-1053
115. Söderman, O., Walderhaug, H., Henriksson, U. & Stilbs, P. (1985) J. Phys. Chem. **89**, 3693-3697
116. Eriksson, P.-O., Lindblom, G. & Arvidson, G. (1987) J. Phys. Chem. **91**, 846-853
117. Johansson, L. B.-Å. & Söderman, O. (1987) J. Phys. Chem. **91**, 7575-7578
118. Burns, J. L., Cohen, Y. & Talmon, Y. (1990) J. Phys. Chem. **94**, 5308-5312
119. Reiss-Husson, F. (1967) J. Mol. Biol. **25**, 363-382
120. Arvidson, G., Brentel, I., Khan, A., Lindblom, G. & Fontell, K. (1985) Eur. J. Biochem. **152**, 753-759
121. Balmbra, R. R., Clunie, J. S. & Goodman, J. F. (1969) Nature (London) **222**, 1159-1160
122. Tardieu, A. &. Luzzati, V. (1970) Biochim. Biophys. Acta **219**, 11-17
123. Charvolin, J. & Sadoc, J. F. (1990) Colloid Polym. Sci. **268**, 190-195
124. Cribier, S., Bourdieu, L., Vargas, R., Gulik, A. & Luzzati, V. (1990) J. Phys. Colloq. **51**, 105-108
125. Portmann, M., Landau, E. M. & Luisi, P. L. (1991) J. Phys. Chem. **95**, 8437-8440
126. Lutton, E. S. (1965) J. Am. Oil Chem. Soc. **42**, 1068-1070
127. Lindblom, G., Larsson, K., Johansson, L., Fontell, K. & Forsén, S. (1979) J. Am. Chem. Soc. **101**, 5465-5470
128. Anderson, D. M. & Wennerström, H. (1990) J. Phys. Chem. **94**, 8683-8694
129. Lindblom, G., Johansson, L. B.-Å. & Arvidson, G. (1981) Biochemistry **20**, 2204-2207
130. Senak, L., Davies, M. A. & Mendelsohn, R. (1991) J. Phys. Chem. **95**, 2565-2571
131. Mantsch, H. H. & McElhaney, R. N. (1991) Chem. Phys. Lipids **57**, 213-226
132. Holmgren, A., Johansson, L. B.-Å. & Lindblom, G. (1987) J. Phys. Chem. **91**, 5298-5301
133. Holmgren, A., Lindblom, G. & Johansson, L. B.-Å. (1988) J. Phys. Chem. **92**, 5639-5642
134. Nilsson, A., Holmgren, A. & Lindblom, G. (1991) Biochemistry **30**, 2126-2133

The interaction of membrane-intrinsic proteins with phospholipids

Juan C. Gómez-Fernández and José Villalaín

Departamento de Bioquímica y Biología Molecular A, Edificio de Veterinaria, Universidad de Murcia, E-30071 Murcia, Spain

It is a great pleasure for us to be able to make a contribution to this book in recognition of Professor Dennis Chapman, F.R.S., to whom we are greatly indebted for the many things that we have learned working with him and mainly for his friendship.

Here, we will review a topic to which Professor Chapman, together with his many colleagues, has made outstanding contributions.

Intrinsic protein–lipid interactions

During the sixties, the scientific community accepted as a useful working model of membrane structure that many proteins were embedded in a bilayer lipid matrix. The question that immediately arose was the degree to which these proteins, called 'intrinsic', can perturb lipid organization and dynamics and, conversely, how lipids may modulate protein structure and dynamics.

Dennis Chapman was a pioneer in asking these questions [1] and in making important contributions to this field (see [2, 3] for reviews).

It must be said in the first place that some methodological problems are inherent in many of the studies performed in this field. A substantial number of these studies have been carried out with reconstituted samples, and this may introduce various problems such as lack of similarity with biological membranes, contamination with detergents or organic solvents, and poor reproducibility.

As stated previously [3] some aspects that are to be taken into account when examining lipid–protein systems include the following: (i) the fluidity status of the sample, i.e. whether the temperature of the experiment is above or below the onset temperature of the gel to liquid-crystalline phase transition (T_c) of the system; (ii) the lateral organization of the membrane, where proteins may be aggregated, or heterogeneously distributed among different domains; and (iii) the molar ratio of lipid to protein.

Macroscopic studies

The effect of intrinsic proteins on the organization of the membrane has been addressed using a number of techniques, such as freeze-fracture electron microscopy, differential scanning calorimetry (DSC), n.m.r. and, only in a very few cases, X-ray diffraction. It is obvious that when studying phase transitions using 'microscopic techniques' such as n.m.r., e.s.r. or vibrational spectroscopy, we are also using a 'macroscopic approach' in a broad sense; we will therefore include some of these studies under the heading of 'phase transitions'.

The principal conclusions obtained from these techniques will now be summarized.

Freeze-fracture electron microscopy

In most studies of systems formed by intrinsic proteins reconstituted with synthetic lipids, it has been observed that whereas proteins are distributed randomly in the plane of the bilayer at temperatures above T_c, phase separations leading to protein patches are observed below T_c, triggered by lipid crystallization.

This general behaviour has been found in (Ca^{2+}, Mg^{2+})-ATPase from sarcoplasmic reticulum [4, 5], bacteriorhodopsin from the purple membrane of *Halobacterium halobium* [6-8], lipophilin [9] and acetylcholinesterase from human erythrocytes [10] (see Fig. 1). The same has also been observed in biological membranes of micro-organisms like *Acholeplasma laidlawii B*, whose lipids contain equimolar amounts of palmitoyl and myristoyl chains [11], and in *Mycoplasma mycoides* var. *capri* or a strain adapted to grow with low cholesterol concentrations [12]. Phase separations of this type have also been reported in inner mitochondrial membranes [13, 14].

The mechanism underlying this type of protein aggregation has been discussed by Chapman *et al.* [15]. At low protein content, when an intrinsic protein is interpolated into a lipid-crystal bilayer, if it does not fit adequately in a space normally occupied by an integral number of lipid chains, imperfections in the packing will result. If the protein concentration is very low these imperfections might be absorbed by the crystal lattice as dislocations. However, as protein concentration is increased after lipid crystallization, the protein will become trapped within the dislocations that form spontaneously as a result of thermal fluctuations during crystallization. As the protein concentration is increased further, a point will be reached at which a dislocation density in excess of that which occurs spontaneously in the pure lipid is produced, and hence the zone refining of the protein into aggregates could not be prevented by its entrapment in packing faults. Consequently, localized regions with a relatively high protein:lipid ratio will be produced.

On the other hand, it the protein:lipid ratio is increased, a limiting point will be reached at which the lipid phase transition will be smeared out

Fig. I.   Freeze-fracture electron microscopy of a DPPC–bacteriorhodopsin recombinant (31:1 molar ratio) quenched from above (a) and below (b) the T_c of the lipid

 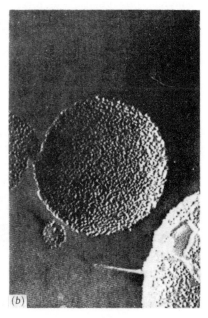

Note the crystalline patches of protein which occur within the lipid bilayer matrix. Magnification ×98 000. Reprinted from [112] with permission.

and removed, and hence freeze-fracture electron microscopy will not reveal a clear phase separation, so that samples thermally quenched from above and below T_c will have a similar lateral organization.

However, protein aggregation above the phase transition has been reported by Lentz et al. [16] for reconstituted (Ca^{2+}, Mg^{2+})-ATPase from sarcoplasmic reticulum. These authors observed protein aggregation below the phase transition and, following the general behaviour discussed above, only at high protein concentration was phase separation of protein-rich patches not observed. The reasons why protein also aggregated above the phase transitions are not clear, as this protein has not been observed to aggregate by other authors, either in reconstituted systems [4, 5] or in native sarcoplasmic reticulum studied by freeze-fracture and other electron microscopic techniques [17].

Phase-transitions

DSC is a very informative technique about the thermodynamic parameters associated with thermally induced phase transitions and the pertubation enforced on them by intrinsic proteins. The reader is referred to a review published recently on this field [18] for detailed analysis of these topics.

From the study of reconstituted systems containing intrinsic transmembrane proteins, such as myelin proteolipid protein [19-22], bacteriorhodopsin [6, 7], cytochrome oxidase [23-25], cytochrome b_5 [26] and (Ca^{2+}, Mg^{2+})-ATPase [5, 27] (see Fig. 2), it can be concluded that all of these systems reduce the cooperativity of gel to liquid-crystalline phase transitions. The pretransition, when the lipid employed displays it, is abolished at relatively low protein concentration. The phase transition is slightly broadened at high protein concentration, although the midpoint temperature of the gel to liquid-crystalline phase transition (T_m) does not change very much, and the enthalpy change (ΔH) decreases linearly in many cases.

The phase transitions of reconstituted systems of the intrinsic proteins mentioned above have also been studied using spectroscopic techniques such as fluorescence depolarization of the probe diphenylhexatriene (DPH) [5, 16, 22, 24, 28, 29], Fourier transform i.r. [30-32] and e.s.r. [33]. The validity of these techniques has been reviewed in detail [18].

Although the behaviour described above can be said to be general, some peculiarities of the different systems and discrepancies of the results or interpretations reported by various authors will now be discussed for some representative intrinsic proteins.

One study on myelin proteolipid apoprotein [34] reported a high temperature shoulder and a shift of the main transition to higher temperatures. However, this was not reproduced by other authors [19-22].

Slight shifts of the main phase transition to lower temperatures were reported for bacteriorhodopsin [6, 7] and small shoulders located at lower temperatures were found for (Ca^{2+}, Mg^{2+})-ATPase [16, 30]. In the case of cytochrome b_5, it has been found that at high protein concentrations two superimposed peaks arise [26], one of them sharp and centered at the same T_c as the pure lipid, which was attributed to pure lipid, and the second and broader peak located at slightly lower temperatures, corresponding to lipid affected by the presence of protein.

A first point to be discussed is the linearity of the decrease of ΔH when plotted against protein concentration. It has been observed in some cases that, at high concentrations of protein, a non-linear relationship appears so that from a certain point the decrease in enthalpy falls. This was first observed by Chapman *et al.* [35] for gramicidin A and it was interpreted as being a result of polypeptide aggregation.

Fig. 2. **The DSC calorimetric heating curves for pure dimyristoylphosphatidylcholine (DMPC) and DMPC-ATPase systems**

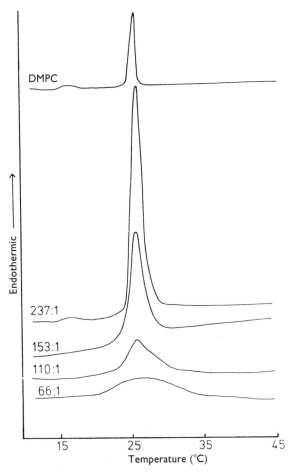

Molar lipid:protein ratios are indicated on the curves. Reprinted from [27] with permission.

Similar results were also obtained for the myelin proteolipid protein [19, 22]; this was in contrast to the findings of other authors who noted a linear relationship [21]. Protein aggregation at high protein concentration could also explain the lack of linearity in the decrease of ΔH.

The procedure used for reconstruction is something that should be considered in order to understand these discrepancies. In contrast with previous results [5, 27], a lack of linearity in the decrease of ΔH, as the protein concentration was increased, was also observed for the (Ca^{2+}, Mg^{2+})-ATPase by Lentz et al. [16]. At high protein concentrations, the effect of protein increase on decreasing ΔH was less [16]. It should be noted that, using freeze-fracture electron microscopy, these authors also observed the peculiar patterns of protein disposition discussed above, with protein aggregation, even at temperatures higher than T_c.

Another point to be considered regarding the quantitative effect of proteins on ΔH is the calculation made by many authors, which involves extrapolating the linear part of the plots of ΔH versus protein concentration to obtain the abscissa value for $\Delta H = 0$. As a general trend, the number of lipid molecules removed from the transition per molecule of protein depends on the size of the protein. However, the scatter of the data found in the literature (see [18] for a review) makes it difficult to arrive at a proportionality, even when considering only phosphatidylcholines. The same problem of data scattering arises if an attempt is made to demonstrate a dependency on the length of the hydrocarbon chains.

The abcissa value for $\Delta H = 0$ is often taken as an indication of the number of molecules of lipid removed from the transition by each molecule of protein. Some authors have interpreted these data as indicating the existence of a special 'boundary lipid' layer separating it from the bulk lipid phase. Upon cooling, the protein will segregate laterally but with its 'boundary lipid', the latter not participating in the thermodynamic phase transition and therefore decreasing the apparent ΔH of the transition. However, this picture may not explain several important facts. Firstly, in reconstituted systems above T_c, it is known that at least some proteins like bacteriorhodopsin [36] or (Ca^{2+}, Mg^{2+})-ATPase, which has been suggested to be a dimer in the membrane [37, 38], may establish protein–protein contacts, and therefore the number of lipids which contact each protein is not constant, but decreases with decreasing lipid:protein ratio. Secondly, below T_c, extensive protein aggregation occurs as noted for bacteriorhodopsin [6–8], (Ca^{2+}, Mg^{2+})-ATPase [4, 5] and myelin proteolipid protein [9], and this may lead to further protein–protein contacts. Thirdly, electron diffraction studies of the native purple membrane of bacteriorhodopsin have shown that, in addition to the lipid in direct contact with the protein, there is a population of phospholipid molelcules filling the 'holes' created by the protein trimers in their hexagonal arrangement [36].

The proposed explanation [5, 7] for the reduction in ΔH observed in intrinsic protein/lipid systems, occurring as a function of protein concentration, is that, as the lipid crystallizes below its main T_c, proteins are excluded from the crystallizing lipid. The proteins in the form of aggregates take with them solvating lipid and form a high number of protein/lipid patches. The lipids in these patches make no contribution to the cooperative melting of the lipid chains or to the enthalpy. The remaining enthalpy would arise from the residual pure lipid region from which the protein has been excluded. As more protein is included in the lipid bilayer,

Fig. 3. **Densitometer tracings of a temperature series of wide-angle X-ray diffraction patterns of Ca^{2+}-ATPase recombinants with lipid:protein molar ratios of 93:1 (*a*) and 53:1 (*b*)**

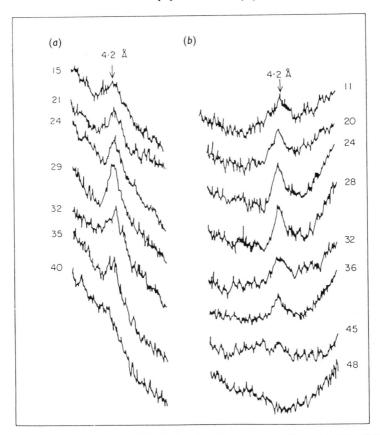

Reprinted from [39] with permission.

this patch process continues. The reduction in the number of lipid molecules available to form the pure lipid region causes a reduction in its cooperativity and also a reduction in the enthalpy of the melting process.

On the basis of research into (Ca^{2+}, Mg^{2+})-ATPase [5, 39], theoretical studies have been carried out [40]. As well as essentially supporting the explanation given above, this work helped to explain the results of wide-angle X-ray diffraction [39] and ^{2}H-n.m.r. studies [41]. It should be said that X-ray diffraction is an essential technique for establishing the structure of lipids in membranes. However, it has been used only in a few cases for studying protein effects on lipid phase behaviour (Fig. 3). Hoffmann *et al.*

Fig. 4. **Approximate phase diagram for a DPPC–protein ($M_r \approx 15\,000$) bilayer**

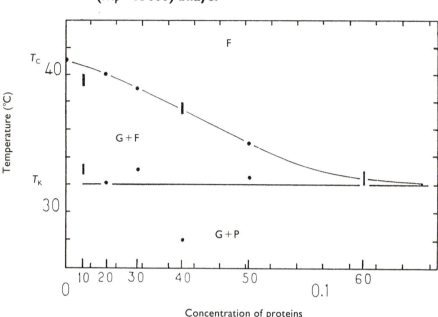

F denotes a fluid phase of homogeneously distributed proteins in a bilayer of predominantly melted lipids, G denotes an ordered essentially pure lipid phase with the chains extended and P denotes the protein-rich phase. The number of proteins is related to their concentration as shown. $T_c = 41°C$ is the main melting temperature of a bilayer of pure DPPC while T_k ($\approx 32°C$) is the temperature at which the protein-rich 'cluster' begins to melt. The replacement of dots by vertical lines at 10, 40 and 60 proteins indicates uncertainty in the temperature of the process in question. Reprinted from [40] with permission.

[39] showed that the amount of crystalline lipid, as monitored by the 0.42 nm reflection of dipalmitoylphosphatidylcholine (DPPC)-ATPase samples, began to decrease at temperatures of about 32 °C, i.e., well below the main T_c of pure DPPC. The computer simulation studies of Lookman et al. [40] supported the explanation given previously for this and other observations [5, 39, 41] elaborating a phase diagram (Fig. 4). In this phase diagram, a two-phase distribution was deduced, composed of a protein-rich cluster and an essentially pure lipid phase at temperatures above 32°C, i.e., the temperature at which the crystalline lipid began to melt. The protein-rich region contained a minimum number of lipids, all of which remained in their melted state, on average, down to temperatures above 24°C. At temperatures below T_c but above 32°C, an essentially pure lipid

phase coexists with the protein-rich cluster, which gradually comes apart, thereby melting as the temperature increases. It is deduced that this could be brought about by melting lipids on the periphery of the cluster, which allows the proteins to diffuse apart. The lipid melting would increase rapidly at some temperature between 32°C and T_c, when the remainder of the lipids melt. For $T > T_c$, all the lipids would melt, on average, and the proteins would diffuse more rapidly and over some time interval would display a random distribution.

The melting observed at about 30 °C for DPPC–(Ca^{2+}, Mg^{2+})-ATPase at an 80:1 molar ratio was also reported using i.r. spectroscopy [30], and closely agrees with an increase of enzymic activity [33] and protein rotational diffusion [39]. Additionally, using high-sensitivity DSC, Lentz et al. [16] observed transitions taking place at temperatures lower than that of the pure phospholipid.

Within this model, and assuming a random disposition of lipids and proteins, it is difficult to know the amount of lipid that is going to be trapped by intrinsic proteins when they are phase separated, although it seems fair to assume that as the protein molecules are of a larger size, more lipid molecules will be necessary to fill the spaces left between protein molecules and more lipid molecules are likely to be squeezed between protein molecules. Similarly, the lipid surrounding these protein/lipid patches, which will melt before the bulk lipid, as proposed by Lookman et al. [40], will depend on the size of the patches which, in turn, depends on the size of the protein molecules.

Other models have been developed to account for the experimental results obtained with lipid/intrinsic protein systems. Mouritsen and Bloom [42] proposed a model of hydrophobic mismatch, the so-called 'mattress' model. In this model, the physical mismatch between the *trans*-bilayer region of the protein and the pure lipid bilayer thickness will result in distortion of the bilayer, which will influence the lateral mixing of proteins and lipids. Proteins with small hydrophobic dimensions will stabilize the thin fluid phase whereas those with hydrophobic dimensions greater than the lipid bilayer thickness would stabilize the thicker gel phase. However, this model does not adequately predict the phase separation of proteins, which takes place below the phase transition at relatively low protein concentrations and which has been detected using freeze-fracture electron microscopy for a number of proteins (see above).

In the same direction, Morrow et al. [43] proposed another model in which the solute concentration dependence of the transition enthalpy in lipid/protein systems is interpreted in terms of regular solution theory. Regions of two-phase equilibrium may exist in the system and the protein would not remove lipid from the main transition but would instead influence the transition thermodynamics for the bilayer as a whole. This model does not need to postulate the existence of any boundary lipid or special lipid, i.e., only one type of lipid would exist as shown in n.m.r. (see below).

As discussed by Scott [44], the preference of the protein for fluid bilayer phase can lead to domain formation and phase separation.

It would be interesting to see if a model that could take into account regions of two-phase equilibrium, plus protein exclusion from crystallized lipid at temperatures below T_c and at relatively low protein concentrations, could accurately predict experimental results.

Before finishing this section, we would like to comment briefly on a class of glycoproteins, which are intrinsic because of a part of them is embedded in the bilayer matrix, but have much of their mass outside the membrane. This is the case of glycophorin which has been analysed in great detail and for which a model has been put forward [45]. In this case a protein molecule may affect many lipid molecules through polar interactions. Depending on protein concentration, the conformation adopted by these proteins will vary and this conformation will be very important in determining lipid–protein interactions.

Microscopic effects

We will now focus our attention on how intrinsic proteins affect lipid dynamics and order, as seen by different spectroscopic techniques.

Anisotropic motion of spin-probes as seen by e.s.r.

E.s.r. was one of the first techniques to be introduced for the study of lipid–protein interactions [46]. It relies on the use of probes, most frequently nitroxide probes, bound to free fatty acids or to phospholipids.

However, some problems inherent to the use of e.s.r. probes have to be addressed before the results can be interpreted. Firstly, do spin labels behave exactly as membrane lipids? There are some reports suggesting that this is not the case. It was found that the nitroxide group, when attached to some hydrocarbon chains, tends to be localized near the lipid–water interface because of its polarity [47]. Note that if the nitroxide group tends to 'float' in the membrane, the size and the shape of this acyl chain will be peculiar, and certainly not identical to an unlabelled one. It is not surprising, therefore, that spin labels have been found to introduce perturbations in lipid monolayers [48] and that some spin labels have been found to cluster even in fluid membranes [49, 50]. It is thus very likely that nitroxide spin probes will not uniformly distribute in a lipid/protein system. A second question to address is whether spin probes interact specifically with the proteins? This was suggested to occur through hydrogen bonding [51], although later results appear to rule out this explanation [52, 53]. Nevertheless this possibility should be checked for each spin label–protein pair.

A considerable number of systems, including biological membranes and proteins reconstituted with defined lipids, have been studied by e.s.r. (see [54, 55] for reviews). The general outcome of these experiments is that two components are seen clearly in systems below the lipid phase transition temperature and also in systems containing a high concentration of protein (Fig. 5). The presence of two components is less evident in systems above the phase transition, particularly if they contain a low protein concentration.

Fig. 5. **E.s.r. spectra of spin-labelled stearic acid (1 molar % of total lipid) as a function of temperature for the system DPPC–ATPase 29:1**

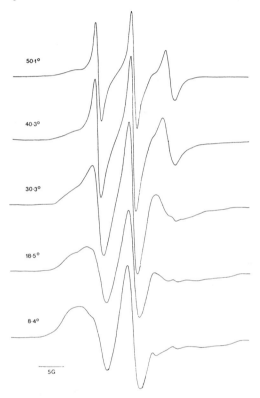

Reprinted from [33] with permission.

The protein effect on probe motion has been analysed in many cases by assuming the line of shape of one component, commonly pure lipid, and subtracting a given amount of that component from the experimental spectrum to obtain the other component. It is clear that this procedure is subject to considerable uncertainty as discussed in [54]. Using this type of approach the fraction of immobilized probe, in the presence of the protein, has been estimated [56].

The interpretation given originally [46] to the appearance of two spectral components in the presence of intrinsic proteins was that two different lipid environments were being sensed by the probe, one corresponding to phospholipid immobilized as a consequence of its contact with the protein, the so-called 'boundary lipid', and the other unaffected or bulk lipid.

The fraction of immobilized probe is often used to estimate the fraction of immobilized lipid which was said to correspond to the number of lipid molecules in direct contact with the protein or 'boundary' layer. This type of calculation presents some problems. One of them is that in many cases proteins have been assumed to be cylindrical rods; this may be an oversimplification. Although there seems to be a certain agreement in some cases between the calculated number and that estimated using spin probes, it is not clear whether this is just fortuitous, as the agreement is not so good in other cases.

Certainly, as discussed above for thermal studies, it is difficult to understand how each protein molecule will be permanently surrounded by a stoichiometric 'annulus' or 'boundary layer' of lipids, since this naive model excludes possibilities such as protein–protein contacts that are nowadays widely accepted to take place, and which will be more likely to occur as the protein concentration is increased. At the same time, some e.s.r. studies [56] are interpreted as suggesting that not only will the layer of lipids in contact with the protein have a restricted motion, but this restriction would extend to a number of layers around each protein. If this is true it is still more difficult to extract stoichiometric conclusions from the fraction of immobilized probes as seen through e.s.r. studies.

Related to that, another interpretation has been given to e.s.r. spectra of lipid/protein systems, showing two components. Equations have been derived that describe changes in the spectrum of e.s.r. as well as in the steady-state fluorescence polarization of the probe DPH as a function of the intrinsic molecule concentration in lipid bilayer membranes [57, 58]. It is supposed in this work that a DPH or a spin-labelled chain is equivalent to an unlabelled lipid hydrocarbon chain with respect to their general space-filling properties. As discussed above this might not be the case. But accepting that, Pink et al. [58] explain e.s.r. spectra and DPH fluorescence polarization results in terms of changes of motion of these probes by their proximity to at least one protein. Making these assumptions they are able to conclude from the analysis of experimental data (see Fig. 6) that there is no evidence for a fixed stoichiometric ratio of 'boundary lipids' (boundary probes) to each protein molecule in the membranes studied, at temperatures above the phase transition, where proteins are distributed randomly in the plane of the membrane. Note that this conclusion relates to probes, and there is no indication of whether or not unlabelled lipids will behave in a similar manner.

The presence of two spectral components when intrinsic proteins are present is also hampered by the fact that it depends on protein concentration and the temperature of the experiment. It was concluded [59] that at low

Fig. 6. **DPH and e.s.r. data plotted as a function of protein concentration**

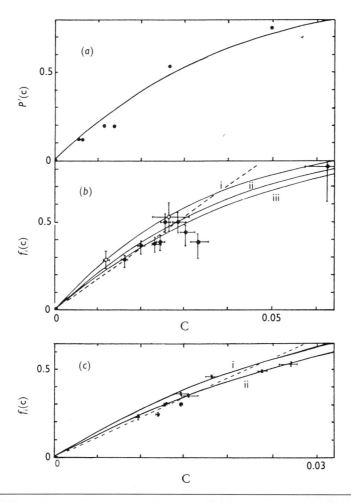

(a) DPH steady-state fluorescence polarization, $P'(c)$, as a function of Ca^{2+}-ATPase concentration, C, in DPPC at 46°C: ●, data points from [5] are replotted here. (b) E.s.r. fraction-immobilized component, $f_i(c)$, as a function of Ca^{2+}-ATPase concentration, C, in the delipidated sarcoplasmic reticulum membranes ●, data points from [113] replotted. The solid lines are plots for dimers with (i) $M = 56$, (ii) $M = 48$ and (iii) $M = 44$; ○, data points from [114]. (c) E.s.r. fraction-immobilized component, $f_i(c)$, as a function of Ca^{2+}-ATPase concentration, C, in egg yolk lecithin bilayers at 25°C: ○, data points from [115] obtained using a proxyl spin-label. The solid lines are plots for dimers with (i) $M = 64$ and (ii) $M = 56$. Reprinted from [58] with permission. Abbreviations used: C, intrinsic molecule concentration; M, maximum number of sites which can be adjacent to a protein.

protein:lipid ratios and at high temperatures there is no immobilized lipid. These observations have been also made by other authors (see for example [60, 61]).

In summary, caution should be exercised when interpreting e.s.r. spectroscopy of membrane probes and when extending these observations to biological membranes.

Lipid-intrinsic protein interaction as seen by n.m.r.
The 'boundary layer' model, as explained above, was extended by later workers [33], who interpreted the immobile component seen in e.s.r. spectroscopy of probes included in (Ca^{2+}, Mg^{2+})-ATPase reconstituted in DPPC as corresponding to a long-lived shell of lipid bound to and surrounding the protein with long lifetimes, in some cases as long as minutes. This was so-called 'annulus-lipid'.

However n.m.r., which was applied to these studies thereafter, did not corroborate the interpretation given to some e.s.r. studies of membrane spin probes. N.m.r. is a well suited technique for studying membrane dynamics, having the advantage over other techniques of not using probes, and thus being non-invasive. ^{2}H-n.m.r. has been especially used in the study of lipid–protein interactions (see [62–64] for reviews). In this case the quadrupole splitting parameter is taken as an indication of the anisotropy of the motion and of the orientation of the axis of motional averaging with respect to the magnetic field.

^{2}H-n.m.r. studies of various reconstituted systems have been reported [41, 65–68] (see Fig. 7) which reveal a number of effects. Firstly, unlike the studies using spin-labelled lipids where two separate signals occur, n.m.r. studies reveal only one signal; i.e. the ^{2}H-n.m.r. studies show that only one homogeneous lipid environment exists in the fluid lipid bilayer. Thus, there is no evidence for stoichiometric complexes of protein and 'boundary' of 'annular' lipids. Secondly, the deuterium quadrupole splitting parameter of the lipid shows only slight changes as a result of the presence of intrinsic proteins in the fluid lipid matrix and, at high protein concentrations, can even show a decrease in the splitting, i.e. an increase in static disorder. Thirdly, 'spin-lattice T_1 relaxation times', which provide information about dynamic order, indicate that the lipid chain segment re-orientation in the presence of the protein is decreased.

As more systems were examined by this technique, confirming the results described above, suggestions were made to reconcile n.m.r. results with some of the interpretations given previously to e.s.r. spin-label studies. One suggestion was that each of these two techniques report on events occurring on different time scales. Whereas the time scale of the e.s.r. measurements is about 10^{-8} s, that of ^{2}H-n.m.r. is about 10^{-5} s. This might mean that if a perturbation occurs to a probe on a time scale of, for example, 10^{-7} s, it will be visualized on the e.s.r. but not the ^{2}H-n.m.r. spectrum. Note that the time scale of e.s.r. is very close to that of the lateral diffusion of lipids

Fig. 7. Deuterium n.m.r. spectra of DMPC-d$_3$, DPPC-d$_6$, and their ATPase (sarcoplasmic reticulum ATPase phosphohydrolase EC 3.6.1.3) complexes as a function of temperature

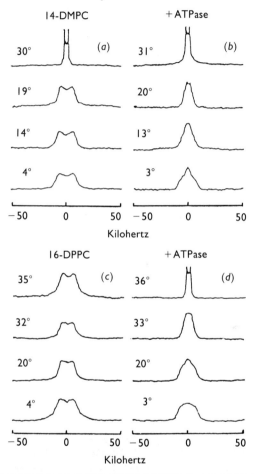

(a) Pure DMPC-d$_3$; (b) DMPC-d$_3$–ATPase 41:1; (c) pure DPPC-d$_6$; and (d) DPPC-d$_6$–ATPase 47:1. Reprinted from [41] with permission. Abbreviations used: DMPC-d$_3$, 1-myristoyl-2-(14,14,14-trideuteriomyristoyl)phosphatidylcholine; DPPC-d$_6$, 1,2-bis-(16,16,16-trideuteriopalmitoyl)phosphatidylcholine.

in the fluid bilayer: the time for a lipid probe to move its own diameter in pure phospholipid bilayers is about 10^{-7} s [69]. This is something of concern for e.s.r. spectroscopic measurements.

After n.m.r. showed that only one type of lipid can be detected in the presence of proteins, some e.s.r. workers suggested the presence of a 'dynamic boundary layer'. This new model means that the exchange between lipids in contact with the protein and those in bulk will be rapid so that n.m.r. will not distinguish between 'boundary' and 'free' lipid. In other words, a molecule may appear to be immobile on a short time scale and yet mobile on a longer time scale.

^{2}H-n.m.r. is able to detect differences between the order and dynamics of the lipids in the gel state and those in the fluid condition [41] (see Fig. 7). The disorder introduced in lipids in the gel state is explained by protein segregation into protein-rich phases where trapped lipid will not be found in a crystalline state.

Apart from deuterium, other nuclei such as ^{13}C and ^{31}P have also been used to study the relaxation mechanism of lipids in the presence of intrinsic proteins [41, 70]. In general, these studies suggest that the polar head group of the phospholipids is affected by the presence of the protein, but only slightly, and that these effects are sensed by all the phospholipids in the membrane, so that again only one phospholipid population is found.

Fluorescence depolarization of the DPH probe

Steady-state dynamic studies using the DPH probe have been carried out on proteins reconstituted with lipids and in biological membranes (reviewed in [54]).

In steady-state measurements, it is supposed that the probe is reporting a static order of the membrane, but again some questionable assumptions have to be made, such as supposing that the DPH molecule behaves exactly as a membrane lipid molecule. This is probably not true because the rigid DPH molecule cannot have the flexibility and reorientation capability of an acyl chain, and may not be sensitive to small disordering of the CH_2 groups [57].

When this probe was used to monitor the effect of intrinsic probes on membranes, it was found that at temperatures above T_c the steady-state polarization parameter increased to reach a maximum saturation value. Therefore, this probe apparently indicates that the order of the lipid system was increased by the presence of the intrinsic protein. This result was in contradiction with n.m.r. findings, as described above, which indicated a slight ordering of the lipid, at least at moderate and high protein concentrations.

A similar explanation to that described above for e.s.r. has been suggested for this observation [57, 58]. With this interpretation, when proteins are present in the lipid bilayer, the motion of the rigid probe is affected mainly by collisions with these intrinsic proteins. As the protein concentration is increased, the polarization value is also increased until a saturation effect is seen [5, 16, 27, 57] (see Fig. 6). This can be understood if it is considered that as the protein concentration increases, protein–protein contact becomes increasingly probable. Therefore, the maximum contact allowed between molecule and protein decreases. Hence the number of probes (or lipids?) influenced by an intrinsic protein will not be constant but will depend on the protein concentration in the lipid matrix.

Vibrational spectroscopic studies
Both Raman and i.r. spectroscopy are non-invasive techniques that have been used to study microscopic effects of intrinsic proteins on membrane lipids. Because they have time-scales of the order of 10^{12} s and they are free from artifacts, such as vesicle tumbling, that can be encountered using n.m.r. they are a very good complement to the latter.

These techniques consistently reveal that intrinsic proteins do not appreciably change the proportion of *gauche* isomers of the acyl chains at temperatures above T_c. This has been found for reconstituted systems of myelin proteolipid apoprotein using Raman spectroscopy [71], for gramicidin A, (Ca^{2+}, Mg^{2+})-ATPase (see Fig. 8) and bacteriorhodopsin using i.r. [30], and also for sarcoplasmic reticulum membranes compared with pure lipid vesicles and for reconstituted (Ca^{2+}, Mg^{2+})-ATPase using Fourier transform i.r. [72].

However, at temperatures below T_c, an increase in the proportion of *gauche* isomers was reported for a number of intrinsic proteins [30]. This was interpreted in terms of protein aggregation forming protein-rich patches, where lipids are not expected to be in the crystalline state.

Specificity of liquid-protein interactions

Physical studies of lipid-protein contacts
The first physical techniques used included the monitoring of lipid-protein contacts in a lipid mixture, in which a given lipid is specifically labelled by means of attached spin labels [73] or brominated acyl chains [74] (see [54] and references therein for a review). Another approach has been to observe a preference of partitioning of the protein through DSC [23, 75] or Fourier transform i.r. [76, 77] when it is reconstituted in a binary lipid mixture where both phospholipids have well separated phase transitions.

The main specificity that can be appreciated in the contact of lipids to some proteins, such as cytochrome c oxidase [23] and (Na^+, K^+)-ATPase [78] is that these proteins seem to show a certain preference for negatively charged lipids such as cardiolipin or phosphatidylserine, with respect to phosphatidylcholine. However, other proteins like (Ca^{2+}, Mg^{2+})-ATPase, bacteriorhodopsin or rhodopsin do not seem to show any specificity for a given phospholipid. Perhaps polar interactions, through charged groups, may be a key factor for determining specific interactions.

Effects of lipids on protein activity and conformation
It is clear that some proteins must show a high specificity for certain lipids. This is true in the case of lipids acting as enzymatic cofactors, an example

being the association of phosphatidylcholine with D-β-hydroxybutyrate dehydrogenase [79, 80]. In addition, some enzymes that are intrinsic components of membranes are implicated in lipid metabolism, and they must specifically recognize their lipid substrates. Other examples are quinones which act

Fig. 8. Temperature dependence of the maximum wavenumber of the CH_2 asymmetric stretching vibration in (*a*) DPPC–Ca^{2+}-ATPase and (*b*) DPPC–bacteriorhodopsin at the molar ratios indicated

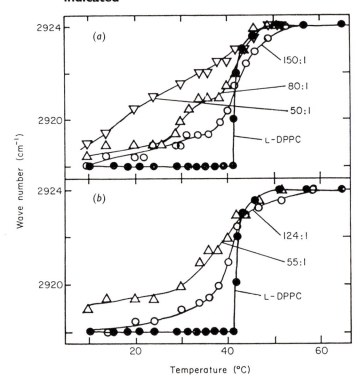

The temperature dependence for pure lipids (●) is also given. Reprinted from [30] with permission.

as components of electron chains. Equally important are lipid mediators which bind to specific membrane receptors like the platelet-activating factor [81] and prostaglandins, thromboxanes and prostacyclins [82].

Besides specificity for certain proteins, lipid composition will determine the physical properties of the membrane and, consequently, may modulate enzymatic activity. Some of these properties are discussed below.

Fluidity

It is generally accepted that membrane fluidity is an important factor in modulating membrane enzymes. This is shown clearly by the great activation which is triggered in reconstituted enzymes, upon passing through the phase transition. Examples are (Ca^{2+}, Mg^{2+})-ATPase from sarcoplasmic reticulum [33, 39], the (Na^+, K^+)-ATPase [83] and β-hydroxybutyrate dehydrogenase [84]. Lipid fluidity has also been shown to directly modulate the overall protein rotational mobility of the (Ca^{2+}, Mg^{2+})-ATPase [85], and the conformation of this protein can also be affected by the fluid status of the membrane [86, 87]. Nevertheless, it is difficult to establish a direct proportionality between fluidity as measured by e.s.r. probes and enzymatic activity [88]. It seems that this only shows that once the protein senses a minimum of fluidity, there are other limiting factors inherent to the catalytic process or to the structure of the membrane.

The effect of cholesterol on membrane proteins has also been widely assumed to be directly related to the regulation of membrane fluidity. Very different results have been obtained for some enzymes, and the discrepancies have not yet been settled in some cases (see [89] for a review). For example, (Ca^{2+}, Mg^{2+})-ATPase has been said to be progressively inhibited by increasing chlolesterol concentration in the membrane [90, 91] whereas, in other cases, cholesterol has been seen to have no effect on the activity of this enzyme [92, 93]. Similarly, with the (Na^+, K^+)-ATPase, there are several reports stating that cholesterol will inhibit the activity of this enzyme ([94, 95], but see [96]). Other transporters which are inhibited by cholesterol in the membrane are the furosemide-sensitive Na^+-K^+-$2Cl^-$ cotransport system [97] and band-3 anion exchanger [98–100]. In other cases cholesterol has been suggested to directly bind to proteins acting as a cofactor [101].

Thickness of the membrane

Membrane thickness is obviously an important factor as, from the thermodynamic point of view, it is necessary for the lipid matrix to solvate adequately the hydrophobic part of the protein. It has been shown, for example for the (Ca^{2+}, Mg^{2+})-ATPase [74, 102, 103] and for the (Na^+, K^+)-ATPase [104, 105], that phospholipids having acyl chains of 16–20 carbons and with a single double bond produce the highest enzymatic activities. It is therefore clear that lipids must be able to provide the protein with a membrane of adequate thickness in order to keep it active.

Electrostatic properties of the lipid–water interface

A number of enzymes, such as the (Na^+, K^+)-ATPase from erythrocytes [106, 107] or cytochrome oxidase from the inner mitochondrial membrane [108], seem to be activated preferentially by anionic phospholipids, although for cytochrome oxidase one report casts doubt on that [109]. This can be thought to be because of polar interactions with charged amino-acyl residues of the proteins. These polar interactions have not been shown, in any of the cases described, to be specific; i.e., there is not a single case in which an identified amino acid has been shown to be bound to a phospholipid by electrostatic interaction. A certain electrostatic potential at the surface of the membrane may be thought to be important in modulating the binding of the substrate. This has been suggested for the ADP–ATP exchanger from the inner mitochondrial membrane [110, 111] which is inhibited by anionic phospholipids, perhaps by electrostatic repulsion with the anionic substrate.

Conclusions

The most important aspects of lipid-intrinsic protein interactions are summarized in the following paragraphs. It should be stressed once again that we are discussing only intrinsic proteins and excluding those glycoproteins which, although anchored in the membrane, have important lipid–protein interactions through residues situated outside of the membrane, as is the case with glycophorin.

Systems at temperatures above T_c and low protein concentration

Proteins are to be found distributed randomly in membranes. Vibrational spectroscopy shows that the proportion of *gauche* isomers is only slightly modified with respect to the pure lipid. Some static ordering may be introduced (as seen for gramicidin A) in the acyl chains. Only some e.s.r. studies of spin-probes have been interpreted as indicating the presence of lipids motionally inhibited by proteins. Even assuming that this interpretation is correct (which is not agreed upon by all authors because the proportion of immobile component is low), it is difficult to say without a high probability of error that the component exists. Thus, the picture is that very slight perturbation is introduced into the system by the presence of low concentrations of proteins (Fig. 9a).

Systems at temperatures above T_c and moderate to high protein concentration

Biological membranes which are fluid and have moderate to high protein concentration represent the most common physiological situation. As the protein concentration is increased, protein–protein contacts are going to become more probable. Vibrational spectroscopic techniques do not indicate a change in the proportion of *gauche* isomers. N.m.r. reveals a slight disorder of the chains that is either caused by squeezing of chains between proteins, or is an artifact of the rough surface of the proteins. Probe techniques like e.s.r. and DPH fluorescence show an increase in the 'immobile probe fraction' and in the polarization of fluorescence, respectively. The e.s.r. result has been interpreted as meaning an immobilization of lipids in contact with the protein or even in other immediate layers. However, e.s.r. and DPH fluorescence results have also been interpreted as meaning probe–protein contacts

Fig. 9. Visualization of some different situations that can be found in protein/lipid systems

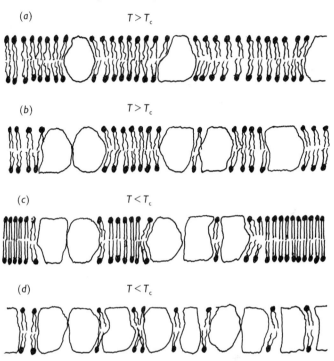

(a) Low protein concentration at a temperature above the phase transition. Some disordered lipid molecules are shown in contact with protein molecules. (b) Moderate to high protein concentration above the phase transition. More disordered lipid molecules than in (a) can be seen here. (c) Low protein concentration at a temperature below the phase transition. Two areas of protein aggregates are shown, together with other areas of pure crystalline lipid. (d) High protein concentration at a temperature below the phase transition. No pure crystalline lipid is present here.

which are more likely to occur as protein concentration is increased. There is no available evidence whatsoever that membrane lipids experience the same protein effects as the probes.

Because protein–protein contacts occur, there is no evidence to support a stoichiometric 'annulus' or 'boundary' layer permanently bound to intrinsic proteins, even if they are in dynamic exchange with bulk lipids.

Thus, the emerging picture is that no important physical effects on lipids exerted by proteins have been found so far, except for those seen through probes, i.e., e.s.r. of spin labels and fluorescence depolarization of DPH (Fig. 9b).

Systems at temperatures below T_c and low protein concentration
Intrinsic proteins are segragated out of the crystalline lipid. In the protein-rich patches formed some lipid will be trapped and will not participate in the phase transition, so that ΔH will decrease. A phase transition may occur at temperatures below that of the pure lipid, corresponding to crystalline lipid (giving a 0.42 nm reflection in X-ray diffraction) surrounding the protein patches. An increase in the proportion of *gauche* isomers is found since perturbed (trapped) lipids are present. Using n.m.r. the same observations are made with respect to lipid disorder and mobility (Fig. 9c).

Systems at temperatures below T_c and high protein concentration
As the protein concentration is increased the crystalline lipid is decreased, until a point is reached at which no free or crystalline lipid remains. At this point ΔH of the transition is zero. Protein–protein contacts are more likely to occur as the protein concentration is increased and no evidence exists for the presence of a stoichiometric lipid 'annulus' surrounding each protein molecule. The proportion of *gauche* isomers will also increase, as well as lipid disorder (Fig. 9d).

Specificity of the interactions and effect on protein activity
Some intrinsic proteins may be able to show specificity with respect to lipids; i.e., they may be able to recognize some lipids in the membrane. This is true in the case of lipid mediators of cell signals, lipids which act as enzyme cofactors, or lipids which are substrates of enzymic reactions taking place in the membrane. But it is important to remark that, for a given protein, these interactions that are going to be very important from the biological point of view will take place only with a minority of membrane lipids present in a biological membrane. On the other hand, it is also quite difficult to reproduce the conditions that will resemble those of biological membranes in reconstituted systems.

The interactions may take place through polar, electrostatic interactions, so that some proteins have a preference for binding anionic phospholipids. Even chlolesterol has been seen to be capable of stabilizing some intrinsic proteins.

Membrane fluidity also seems to be essential, although membrane thickness is even more important, for adequately solvating intrinsic proteins.

We acknowledge support from DGICYT (Spain) through the grants PB87-0704, PB90-297 and PM90-0044. We are also very grateful to Prof. Dennis Chapman for reviewing this manuscript and for his very useful comments. Some of these discussions with Prof. Chapman were possible thanks to NATO Joint Grant 86/0712.

References

1. Chapman, D. (1968) in Biological Membranes: Physical Fact & Function (Chapman, D., ed.), Academic Press, New York
2. Chapman, D., Gómez-Fernández, J. C. & Goñi, F. M. (1979) FEBS Lett. 98, 211–223
3. Chapman, D. (1983) in Membrane Fluidity in Biology (Aloia, R. C., ed.), vol. 2, pp. 5–42, Academic Press, New York and London
4. Kleeman, W. & McConnell, H. M. (1976) Biochim. Biophys. Acta 419, 206–222
5. Gómez-Fernández, J. C., Goñi, F. M., Bach, D., Restall, C. J. & Chapman, D. (1980) Biochim. Biophys. Acta 598, 502–516
6. Heyn, M. P., Blume, A., Rehorek, M. & Dencher, N. A. (1981) Biochemistry 20, 7109–7115
7. Alonso, A., Restall, C. J., Turner, M., Gómez-Fernández, J. C., Goñi, F. M. & Chapman, D. (1982) Biochim. Biophys. Acta 689, 283–289
8. Sternberg, B., Gale, P. & Watts, A. (1989) Biochim. Biophys. Acta 980, 117–126
9. Boggs, J. M. & Moscarello, M. A. (1978) Biochim. Biophys. Acta 515, 1–21
10. Frenkel, E. J., Roelofsen, B., Brodbeck, U., van Deenen, L. L. M. & Ott, P. (1980) Eur. J. Biochem. 109, 377–382
11. Silvius, J. R. & McElhaney, R. N. (1980) Proc. Natl. Acad. Sci. U.S.A. 77, 1255–1259
12. Rottem, S., Yashouv, J., Ne'eman, Z. & Razin, G. (1973) Biochim. Biophys. Acta 323, 495–508
13. Hackenbrock, C. R., Hochli, M. & Chau, R. M. (1976) Biochim. Biophys. Acta 455, 466–484
14. Hochli, M. & Hackenbrock, C. R. (1977) Proc. Natl. Acad. Sci. U.S.A. 73, 1636–1640
15. Chapman, D., Cornell, B. A. & Quinn, P. J. (1977) in Biochemistry of Membrane Transport, FEBS Symposium no. 42 (Semenza, G. & Carafoli, E., eds.), pp. 72–85, Springer-Verlag, Berlin and Heidelberg
16. Lentz, B. R., Clubb, K. W., Alford, D. R., Hochli, M. & Meissner, G. (1985) Biochemistry 24, 433–442
17. Inesi, G. & Kurzmack, M. (1983) in Biomembrane Structure and Function (Chapman, D., ed.), pp. 355–410, MacMillan Press, London
18. McElhaney, R. N. (1986) Biochim. Biophys. Acta 864, 361–421
19. Papahadjopoulos, D., Moscarello, M., Eylar, E. M. & Isac, T. (1975) Biochim. Biophys. Acta 401, 317–335
20. Papahadjopoulos, D., Vail, W. J. & Moscarello, M. (1975) J. Membrane Biol. 22, 143–164
21. Boggs, J. M. & Moscarello, M. A. (1978) Biochemistry 17, 5734–5739
22. Lavialle, F., Gabrielle-Madelmont, C., Petit, J., Ollivon, M. & Alfsen, A. (1985) Biochemistry 19, 6044–6050
23. Semin, B. K., Saraste, M. & Wikstrom, M. (1984) Biochim. Biophys. Acta 769, 15–22
24. Rigell, C. W., De Saussure, C. & Freire, E. (1985) Biochemistry 24, 5638–5646
25. Singer, M. A., Dinda, M., Young, M. & Finegold, L. (1986) Biochim. Cell Biol. 64, 91–98
26. Freire, E., Markello, T., Rigell, C. & Holloway, P. W. (1983) Biochemistry 22, 1675–1680
27. Gómez-Fernández, J. C., Goñi, F. M., Bach, D., Restall, C. J. & Chapman, D. (1979) FEBS Lett. 98, 224–228
28. Heyn, M. P. (1979) FEBS Lett. 108, 359–364
29. Rehorek, M., Dencher, N. A. & Heyn, M. P. (1985) Biochemistry 24, 5980–5988
30. Cortijo, M., Alonso, A., Gómez-Fernández, J. C. & Chapman, D. (1982) J. Mol. Biol. 157, 597–618
31. Anderle, G. & Mendelsohn, R. (1986) Biochemistry 25, 2174–2179
32. Goñi, F. M., Cózar, M., Alonso, A., Durrani, A. A., García-Segura, L. M., Lee, D. C., Monreal, J. & Chapman, D. (1988) Eur. J. Biochem. 174, 641–646
33. Hesketh, T. R., Smith, G. A., Houslay, M. D., McGill, K. A., Birdsall, N. J., Metcalfe, J. C. & Warren, G. B. (1976) Biochemistry 15, 4145–4151
34. Curatolo, W., Sakura, J. D., Small, D. M. & Shipley, G. G. (1977) Biochemistry 16, 2313–2319
35. Chapman, D., Cornell, B. A., Eliasz, A. W. & Perry, A. (1977) J. Mol. Biol. 113, 517–538
36. Cherry, R. J. (1979) Biochim. Biophys. Acta 559, 502–516
37. Moller, J. V., Andersen, J. P. & LeMaire, M. (1982) Mol. Cell. Biochem. 42, 83–107
38. Dux, L. & Martonosi, A. M. (1983) J. Biol. Chem. 258, 2599–2603
39. Hoffmann, W., Sarzala, M. G., Gómez-Fernández, J. C., Goñi, F. M., Restall, C. J., Chapman, D., Heppeler, G. & Kreutz, W. (1980) J. Mol. Biol. 141, 119–132
40. Lookman, T., Pink, D. A., Grundke, E. W., Zuckerman, M. J. & deVerteuil, F. (1982) Biochemistry 21, 5593–5601

41. Rice, D. M., Meadows, M. D., Scheinman, A. O., Goñi, F. M., Gómez-Fernández, J. C., Moscarello, M. A., Chapman, D. & Oldfield, E. (1979) Biochemistry 18, 5893–5903
42. Mouritsen O. G. & Bloom, M. (1984) Biophys. J. 46, 141–153
43. Morrow, M. R., Huschilt, J. C. & Davis, J. M. (1985) Biochemistry 24, 5396–5496
44. Scott, H. L. (1986) Biochemistry 25, 6122–6126
45. McDonald, A. L. & Pink, D. A. (1987) Biochemistry 26, 1909–1917
46. Jost, P. C., Griffith, O. H., Capaldi, R. A. & Venderkooi, G. (1973) Proc. Natl. Acad. Sci. U.S.A. 70, 480–484
47. Ellena, J. F., Archer, S. J., Dominey, R. N., Hill, B. D. & Cafiso, D. S. (1988) Biochim. Biophys. Acta 940, 63–70
48. Cadenhead, D. A. & Müller-Landau, F. (1973) Biochim. Biophys. Acta 307, 279–286
49. Hauser, H., Guyer, W. & Howell, K. (1979) Biochemistry 18, 3285–3291
50. Gordon, L. M., Looney, F. D. & Curtain, C. C. (1987) Biochim. Biophys. Acta 898, 202–213
51. Dehlinger, P. J., Jost, P. C. & Griffith, O. H. (1974) Proc. Natl. Acad. Sci. U.S.A. 71, 2280–2284
52. Rousselet, A., Devaux, P. F. & Wirtz, K. W. (1979) Biochem. Biophys. Res. Commun. 90, 871–877
53. Griffith, O. H., Brotherus, J. R. & Jost, P. C. (1982) in Lipid–Protein Interactions (Jost, P. C. & Griffith, O. H., eds.), pp. 225–237, Wiley-Interscience, New York
54. Devaux, P. F. & Seigneuret, M. (1985) Biochim. Biophys. Acta 822, 63–125
55. Marsh, D. (1990) FEBS Lett. 286, 371–375
56. Knowles, P. F., Watts, A. & Marsh, D. (1979) Biochemistry 18, 4480–4487
57. Hoffmann, W., Pink, D. A., Restall, C. J. & Chapman, D. (1981) Eur. J. Biochem. 114, 585–589
58. Pink, D. A., Chapman, D., Laidlaw, D. J. & Wiedmer, T. (1984) Biochemistry 23, 4051–4058
59. Davoust, J., Schoot, B. M. & Devaux, P. F. (1979) Biochemistry 76, 2755–2759
60. Esmann, M., Hideg, K. & Marsh, D. (1988) Biochemistry 27, 3913–3917
61. Li, G., Knowles, P. F., Murphy, D. J. & Marsh, D. (1990) J. Biol. Chem. 265, 16867–16872
62. Oldfield, E., James, N., Kinsey, R., Quintanr, A., Lee, R. W. K., Rothgeb, M., Scharam, S., Skarjune, R., Smith, R. & Tsai, M. D. (1981) Biochem. Soc. Symp. 46, 155–181
63. Seelig, J., Seelig, A. & Tamm, L. (1982) in Lipid–Protein Interactions, vol. 2, (Jost, P. C. & Griffith, O. H., eds.), pp. 127–148, John Wiley & Sons, New York
64. Davis, J. H. (1983) Biochim. Biophys. Acta 737, 117–171
65. Oldfield, E., Meadows, M., Rice, D. M. & Jacobs, R. (1978) Proc. Natl. Acad. Sci. U.S.A. 75, 1616–1619
66. Gally, H. U., Puschka, G., Overath, P. & Seelig, J. (1981) Biochemistry 20, 1826–1831
67. Paddy, M. R., Dahlquist, F. W., Davis, J. H. & Bloom, M. (1981) Biochemistry 20, 3152–3162
68. Tamm, L. & Seelig, J. (1983) Biochemistry 22, 1474–1483
69. Träuble, H. & Sackmann, E. (1972) J. Am. Chem. Soc. 94, 4499–4510
70. Seelig, J. (1983) Biochemistry 22, 1474–1483
71. Lavialle, F. & Levin, I. W. (1980) Biochemistry 19, 6044–6050
72. Mendelsohn, R., Anderle, G., Jaworsky, M., Mantsch, H. H. & Dluhy, R. A. (1984) Biochim. Biophys. Acta 775, 215–224
73. London, E. & Feigenson, G. W. (1981) Biochemistry 20, 1939–1948
74. East, J. M. & Lee, A. G. (1982) Biochemistry 21, 4144–4151
75. Boggs, J. M., Wood, D. D., Moscarello, M. A. & Papahadjopoulos, D. (1977) Biochemistry 16, 2325–2329
76. Jaworski, M. & Mendelsohn, R. (1985) Biochemistry 24, 3422–3428
77. Mendelsohn, R., Brauner, J. W., Faines, L., Mantsch, H. H. & Dluhy, R. A. (1984) Biochim. Biophys. Acta 774, 237–246
78. Zachowski, A. & Devaux, P. F. (1983) FEBS Lett. 163, 245–249
79. Gazzotti, P., Bock, H. G. & Fleischer, S. (1974) Biochem. Biophys. Res. Commun. 58, 309–315
80. Clancy, R. M., McPherson, L. H. & Glaser, M. (1983) Biochemistry 22, 2358–2364
81. Prescott, S. M., Zimmerman, G. A. & McIntyre, T. M. (1990) J. Biol. Chem. 265, 17381–17384
82. Rasmussen, H. (1986) N. Engl. J. Med. 314, 1164–1170
83. Kimelberg, H. K. & Papahadjopoulos, D. (1974) J. Biol. Chem. 249, 1071–1080
84. Houslay, M. D., Warren, G. B., Birdsall, N. J. M. & Metcalfe, J. C. (1975) FEBS Lett. 51, 146–151
85. Squier, T. C., Bigelow, D. J. & Thomas, D. D. (1988) J. Biol. Chem. 263, 9178–9186
86. Gómez-Fernández, J. C., Baena, M. D., Teruel, J. A., Villalaín, J. & Vidal, C. J. (1985) J. Biol. Chem. 260, 7168–7170
87. Gutiérrez-Merino, C. (1987) Arch. Biochem. Biophys. 252, 303–314
88. East, J. M., Jones, O. T., Simmonds, A. C. & Lee, A. G. (1984) J. Biol. Chem. 259, 8070–8071
89. Yeagle, P. L. (1985) Biochim. Biophys. Acta 822, 267–287

90. Warren, G. B., Houslay, M. D., Metcalfe, J. C. & Birdsall, N. J. M. (1975) Nature (London) 255, 684-687
91. Johannson, A., Keighthly, C. A., Smith, B. A. & Metcalfe, J. C. (1981) Biochem. J. 196, 505-511
92. Madden, T. D., Chapman, D. & Quinn, P. J. (1971) Nature (London) 279, 538-541
93. Madden, T. M., King, M. D. & Quinn, P. J. (1981) Biochim. Biophys. Acta 641, 265-269
94. Anner, B. M. (1985) Biochim. Biophys. Acta 822, 319-334
95. Proverbio, F. & Rawlins, F. A. (1978) Mol. Pharmacol. 14, 911-919
96. De Pont, J. J. H. H. M., Peters, W. H. & Bonting, S. L. (1983) Curr. Top. Membr. Transport 19, 163-166
97. Tosteson, D. C. (1981) Fed. Proc. Fed. Am. Soc. Exp. Biol. 40, 1429-1433
98. Gregg, V. A. & Reithmeier, R. A. F. (1983) FEBS Lett. 157, 159-164
99. Grunze, M., Forst, B. & Deuticke, B. (1980) Biochim. Biophys. Acta 600, 860-869
100. Jackson, P. & Morgan, D. B. (1982) Biochim. Biophys. Acta 693, 99-104
101. Artigues, A., Villar, M. T., Fernández, A. M., Ferragut, J. A. & González-Ros, J. M. (1989) Biochim. Biophys. Acta 985, 325-330
102. Caffrey, M. & Feigenson, G. W. (1981) Biochemistry 20, 1949-1961
103. Johansson, A., Keightley, C. A., Smith, G. A., Richards, C. D., Hesketh, T. R. & Metcalfe, J. C. (1981) J. Biol. Chem. 256, 1643-1650
104. Johansson, A., Smith, G. A. & Metcalfe, J. C. (1981) Biochim. Biophys. Acta 641, 416-421
105. Marcus, M. M., Apell, H. J., Roudna, M., Schwendenes, R. A., Weder, H. G. & Läuger, P. (1986) Biochim. Biophys. Acta 854, 270-278
106. Cornelius, F. & Skou, J. C. (1984) Biochim. Biophys. Acta 772, 357-373
107. Roelofsen, B. (1981) Life Sci. 29, 2235-2247
108. Robinson, N. C., Stry, F. & Talbert, L. (1980) Biochemistry 19, 3656-3661
109. Powell, G. L., Knowles, P. F. & Marsh, D. (1985) Biochim. Biophys. Acta 816, 191-194
110. Krämer, R. & Klingenberg, M. (1980) FEBS Lett. 119, 257-260
111. Krämer, R. (1983) Biochim. Biophys. Acta 735, 145-149
112. Chapman, D., Gómez-Fernández, J. C. & Goñi, F. M. (1982) Trends Biochem. Sci. 7, 67-70
113. Jost, P. C. & Griffith, O. H. (1978) in Biomolecular Structure and Function (Agris, P. F., ed.), pp. 25-65, Academic Press, New York
114. Thomas, D. D., Bigelow, D. J., Squier, T. S. & Hidalgo, C. (1982) Biophys. J. 37, 217-225
115. Silvius, J. R., McMillen, D. A., Saley, N. D., Jost, P. C. & Griffith, O. H. (1984) Biochemistry 23, 538-547

Cholesterol–phospholipid interactions and the exchangeability of cholesterol between membranes

Michael C. Phillips

Biochemistry Department, The Medical College of Pennsylvania,
3300 Henry Avenue, Philadelphia,
PA 19129, U.S.A.

Introduction

An aspect of membrane structure that has been of interest to Dennis Chapman over the past quarter of a century is the nature of the cholesterol–phospholipid interaction. As indicated below, together with Stuart Penkett, he made the first direct demonstration of the effects of cholesterol on the phospholipid acyl chains in a bilayer membrane. This seminal study has been confirmed and extended by the application of a host of other biophysical techniques both in Dennis Chapman's laboratory and in other laboratories around the world. This interest in the effects of cholesterol on the structure and function of mammalian membranes has arisen because of the many physiological and pathophysiological consequences of the presence of this sterol in cells.

  This chapter deals first with the structural consequences of the cholesterol–phospholipid interaction. In particular, the ways in which cholesterol modulates the fluidity of membranes will be discussed; however, in the interests of space, the interactions of cholesterol with membrane proteins are excluded. The second topic to be covered is the effects of cholesterol–phospholipid interactions on the kinetics of cholesterol exchange between membranes. By affecting the exchangeability of cholesterol, cholesterol–phospholipid interactions influence cell cholesterol homeostasis. Aberrations in the latter process can have major consequences for the development of atherosclerosis and the incidence of coronary artery disease in humans.

Cholesterol–phospholipid interactions in bilayer membranes

As demonstrated in Fig. 1, Chapman and Penkett [1] used high resolution n.m.r. to show the effects of cholesterol on the acyl-chain motions in small unilamellar vesicles of egg yolk phosphatidylcholine (PC). The differential broadening of the acyl chain resonances at 8–9 p.p.m. relative to the PC polar group $N(CH_3)_3$ resonance at 6.8 p.p.m. is apparent. These data provided the first evidence for a direct interaction between the phospholipid acyl chains and cholesterol molecules. When this study was completed 25 years ago, understanding of the phase behaviour of hydrated PC systems was in its infancy and, in particular, the nature of the PC and cholesterol molecular motions had not been defined. As discussed elsewhere in this volume, Dennis Chapman made many pioneering contributions to answering the above questions. Here, I review those aspects of the cholesterol–phospholipid interaction necessary for understanding the mechanism by which cholesterol is exchanged between membranes.

Chapman and colleagues have used a range of biophysical techniques, including spectroscopy, X-ray diffraction and calorimetry, to study the phase behaviour of hydrated PC systems (for reviews, see [2–5]). In the case of dipalmitoyl PC, at temperatures less than 41 °C the bilayers exist in the gel phase with the acyl chains in a crystalline, all-*trans* conformation. An isothermal melting to a liquid-crystal phase occurs at 41 °C; in this process, the increase in enthalpy permits rotational isomerization to occur in the acyl chains. The 'fluidity' of the PC bilayer is essentially a measure of the proportion of the C—C bonds of the acyl chains forming *gauche* conformers. Fig. 2 summarizes the effects of the gel to liquid-crystal transition on the fluidity of dipalmitoyl PC as a function of the position along the acyl chains [6]. The acyl chains in a gel-phase dipalmitoyl PC bilayer ($T < 41$ °C) are uniformly rigid along their long axis. In liquid-crystal bilayers ($T > 41$ °C), the rigidity, which is inversely proportional to fluidity, is greatest at the polar end of the PC molecule because the glycerylphosphorylcholine groups are closely packed, and this constrains the conformational freedom of the adjacent methylene segments of the acyl chains. The rigidity decreases with increasing carbon number along the acyl chains (Fig. 2).

Cholesterol molecules incorporate readily into hydrated bilayers of PC. At equilibrium, the lamellar phase of dipalmitoyl PC becomes saturated with cholesterol when an equimolar amount of cholesterol is present [7, 8]. The presence of cholesterol molecules dissolved in the PC bilayer has a profound effect on the molecular packing (see Fig. 1); this effect has been studied widely because of its relevance to the function of biological membranes (for reviews, see [2, 9, 10]). The cholesterol molecules are oriented essentially perpendicular to the plane of the bilayer with the hydroxyl group on the C_3 atom hydrogen bonded either to a carbonyl group of a neighbouring PC molecule or to a water molecule. The sidechain of the

Fig. 1. High-resolution proton n.m.r. spectra (60 MHz)

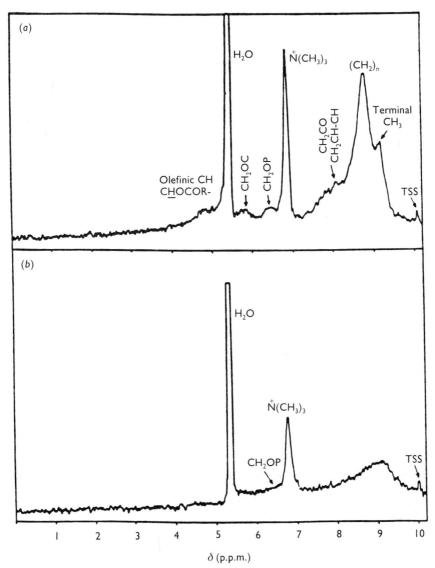

(a) Egg yolk PC vesicles (5% w/w) dispersed in deuterium oxide. (b) A 1:1 molar egg yolk PC:cholesterol mixture dispersed as vesicles in deuterium oxide. Reproduced from [1] with permission. Abbreviation used: TSS, (trimethylsilyl)-1-propane sulphonic acid sodium salt.

cholesterol molecule is located towards the center of the PC bilayer, while the steroid ring system interacts by Van der Waals forces with the hydrocarbon chains of adjacent PC molecules. The rigid ring system interferes with the cooperative lateral packing of the PC acyl chains and reduces the propor-

Fig. 2. **Diagrammatic representation of the variation of fluidity along PC (lecithin) hydrocarbon chains in hydrated dipalmitoyl PC and dipalmitoyl PC/ cholesterol bilayers**

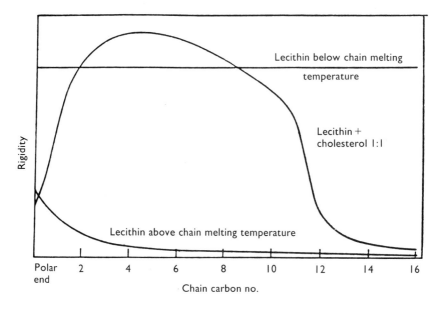

The vertical scale is derived from widths of wide lines in n.m.r. spectra, but is meant to be qualitative only. Reproduced from [6] with permission.

tion of C—C bonds in the *gauche* conformation in liquid-crystal bilayers and reduces the fraction of all-*trans* C—C bonds in gel-phase bilayers. As can be seen from Fig. 2 for bilayers containing a 1:1 molar ratio of cholesterol:PC so that every PC molecule is interacting with a cholesterol molecule, this ordering effect is most pronounced for the first 12 or so carbon atoms along the chain from the polar end of the PC molecule. This disruption of the molecular packing prevents the acyl chains in such a mixed bilayer of cholesterol and dipalmitoyl PC from undergoing isothermal chain melting at 41 °C. Effectively, cholesterol molecules maintain the 12 interacting methylene groups of acyl chains of neighbouring dipalmitoyl PC molecules in a state of fluidity intermediate to the liquid-crystal and gel phases in the temperature range 20–60 °C. When the acyl chains of the PC molecules are more than about 20 °C above their normal melting point, the ordering effect of neighbouring cholesterol molecules is reduced with more methylene groups at the methyl terminal of the chains becoming more mobile [6]. As can be seen from the above, current understanding of the molecular interactions of cholesterol and PC along the axis normal to the bilayer plane is quite well developed. The Van der Waals interactions between cholesterol molecules and the acyl chains of neighboring PC molecules are critical for the solubilization of cholesterol in the bilayer. The polar interactions between cholesterol and PC molecules seem relatively unimportant; there is no strong evidence for direct hydrogen bonding between the two molecular species [5] and the motions of the polar group of PC molecules are not restricted by the presence of cholesterol molecules [1, 6].

The lateral organization along the bilayer plane of the molecules in mixed cholesterol/PC bilayers is not fully understood. The reason for this uncertainty is largely down to the fact that the cholesterol/PC/water phase diagram is complicated and has only been established recently with a reasonable degree of certainty. Magnetic resonance studies have shown that there are situations where cholesterol molecules are not distributed evenly along the bilayer [6, 11, 12]. Hui [13] has reviewed the experimental evidence for lateral phase separation in fully hydrated cholesterol/PC bilayers and discussed the cholesterol-containing domains that form in both model and real biomembranes. Fig. 3 depicts a phase diagram for the hydrated dipalmitoyl PC/cholesterol system which rationalizes most of the experimental evidence; it is apparent that depending upon temperature, lateral phase separation occurs in bilayers containing 5–30 molar% cholesterol. This lateral phase separation has been explained variously in terms of thermodynamic models, models describing the microscopic interactions between cholesterol and PC, and models invoking specific complex formation between cholesterol and PC molecules. Ipsen and colleagues [14] have applied regular solution theory and a microscopic interaction model to derive theoretical phase diagrams similar to that depicted in Fig. 3. The latter model accounts for the gel to liquid-crystal phase transition in pure PC bilayers in terms of acyl-chain degrees of freedom and the PC translational degrees of freedom. In addition, specific interactions between cholesterol and these two sets of PC degrees of freedom are accounted for. It is assumed that cholesterol interacts favorably with PC acyl chains which are in an extended conformation and that it disturbs the translational order in the gel state of PC. The agreement between the resultant theoretical phase diagram (Fig. 3) and experimental data shows that lateral-phase separation in mixed cholesterol/PC bilayers can be explained without invoking the formation of specific PC-cholesterol complexes (but see below).

A major unresolved question about the phase diagram in Fig. 3 is the exact molecular structures of the immiscible phases and the nature of the boundaries between these different domains. A recent ^{2}H n.m.r. study [15]

Fig. 3. Phase diagram for the microscopic model of the hydrated phosphatidylcholine/cholesterol system

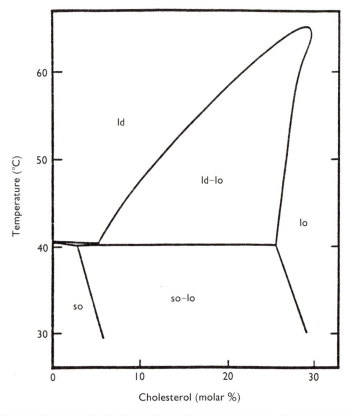

The model parameters are pertinent for dipalmitoyl PC. The gel phase which exists below 41 °C is designated the solid-ordered (so) phase where s refers to the crystalline order and o to the average acyl-chain conformation. The liquid-crystalline phases at low and high molar % cholesterol are called liquid-disordered (ld) and liquid-ordered (lo), respectively. The regions where lateral phase separation occur are designated so-lo and ld-lo. Reproduced from [14] with permission.

has suggested molecular models for the liquid-ordered and liquid-disordered phases. In the liquid-ordered phase, cholesterol molecules effectively extend from one monolayer of the bilayer into the other monolayer (a partially interdigitated packing), whereas in the liquid-disordered phase they span the entire transbilayer hydrophobic region. The fluid phase immiscibility (region ld-lo in Fig. 3) is postulated to be a result of the acyl-chain length mismatch between the flexible phospholipid molecules and the rigid cholesterol molecules, with the phospholipid acyl-chain length being affected by the varying amounts of cholesterol present [15]. Alternative explanations for the lateral phase separation have invoked the formation of specific PC-cholesterol complexes that are only partially miscible with excess PC. Such complexes have been proposed to form in bilayers containing 20, 33 and 50 molar % cholesterol ([16] and references therein). The 1:1 cholesterol:PC complex is thought to be stabilized by a hydrogen bond and the 1:2 cholesterol:PC complex may form by loose association of a second PC molecule. Boundary phospholipid between the domains of 1:2 complexes and free PC is postulated to assume the composition of 20 molar % cholesterol [16]. The fact that it is not necessary to assume stoichiometric cholesterol-PC interactions to derive the theoretical phase diagram in Fig. 3 raises questions about this interpretation. However, the concept of a preferred 1:1 association between cholesterol and PC molecules is consistent with the fact that the maximum equilibrium solubility of cholesterol in fully hydrated PC bilayers is equimolar [7, 8]. At higher levels, a crystalline cholesterol phase forms, so that a vertical line at 50 molar % indicating the 1:1 association has to be added to phase diagrams of the type shown in Fig. 3.

In summary, our understanding of cholesterol-PC molecular interactions is now much more detailed than when Chapman initiated his n.m.r. studies a quarter of a century ago [1], and the conclusion drawn from these initial studies, i.e. that the interactions between PC acyl chains and cholesterol molecules are critical, has been greatly substantiated. Future work should provide more insights into the sizes and structures of the laterally separated domains and the nature of the interfacial regions between them.

Cholesterol exchangeability from different membranes

It is well established that different types of cells growing in culture release cholesterol at different rates to a common extracellular acceptor [17, 18]. As isolated plasma membrane vesicles exhibit the same kinetics for cholesterol efflux as the parent cell, it follows that the structure of the plasma membrane and interactions of cholesterol molecules within the membrane are critical. The differing cholesterol-phospholipid interactions in different types of membranes also influence the distribution of cholesterol within a cell, with the plasma membrane being relatively enriched in cholesterol [19]. The ability of cholesterol to redistribute from the plasma membrane, to either extracellular particles such as high-density lipoprotein or intracellular membranes, is critical for cell cholesterol homeostasis [20].

Mechanism of cholesterol exchange
A convenient system for investigating the mechanism of cholesterol and PC exchange is a dispersion of well characterized, unilamellar cholesterol/PC

vesicles. The kinetics of exchange of radiolabelled lipid are not complicated by the presence of protein, fusion of particles or adsorption to cell membranes. Fig. 4 demonstrates representative time courses for 1-palmitoyl-2-oleoyl PC and cholesterol exchange in such a vesicle system. There are four properties which are characteristic of the kinetics of cholesterol exchange in these systems. First, the rate of cholesterol transfer or exchange is first order with respect to the concentration of cholesterol in the donor vesicle (see inset to Fig. 4). In egg PC vesicles containing 20 molar % cholesterol, essentially all the cholesterol is in a single kinetic pool so that at 37 °C, the half-time ($t_{1/2}$) is about 2 h. This indicates that the rate of transbilayer movement of cholesterol molecules is rapid relative to the rate of transfer from the bilayer. Second, the $t_{1/2}$ of PC transfer is about 25 times longer than that of cholesterol (Fig. 4). Third, the transfer process is strongly temperature-dependent with an experimental activation energy of about 75 kJ/mol (see Fig. 5). Fourth, at high acceptor:donor particle ratios, the rate of transfer is zero order with respect to the concentration of acceptor particles. This indicates that, under this condition, collisions with the aceptor particles are not rate-limiting for cholesterol exchange and that desorption of cholesterol molecules from the donor lipid–water interface (see below) is rate-limiting for the overall transfer process (for reviews, see [17–19]).

By analogy to models for the kinetics of micellization of amphipathic molecules, cholesterol desorption has been analysed in terms of formation of a transition state complex where the cholesterol molecule is attached to the donor vesicle by the tip of its hydrophobic tail (Fig. 6). In this model, a desorbing cholesterol molecule diffuses parallel to its long axis and at right angles to the lipid–water interface. When it has a potential energy equal to the free energy of activation (ΔG^*), it diffuses into the aqueous phase. Any factor that alters ΔG^* will influence the rate of cholesterol desorption. ΔG^* is particularly sensitive to the strength of the interactions of the desorbing cholesterol molecule with its neighbouring phospholipid molecules in the donor lipid–water interface. Cholesterol–phospholipid interactions are altered by the phospholipid composition, cholesterol content, curvature of the interface and the presence of proteins. This chapter focuses on the first two parameters, as reflected in the acyl chain composition, type of phospholipid and level of cholesterol present.

Effects of phospholipid–cholesterol interactions

Changes in the intermolecular interactions of cholesterol in the donor lipid–water interface modify ΔG^* for formation of the transition state (Fig. 6), with the rate-constant for desorption increasing as the interaction energy of cholesterol with its neighbours decreases. As the lateral packing density in cholesterol/phospholipid monolayers and bilayers decreases as the unsaturation of the phospholipid hydrocarbon chains increases [9], the Van der Waals interactions of cholesterol with the phospholipid matrix decrease and desorption is facilitated. The rate constants in Fig. 5 and $t_{1/2}$ values in Table 1

Fig. 4. Relative rates of cholesterol and PC transfer at 37 °C between negatively charged 'donor' unilamellar vesicles and neutral 'acceptor' vesicles

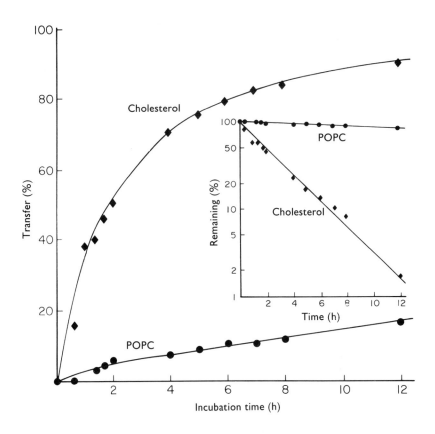

The donor vesicle composition was 20 molar % cholesterol, 65 molar % egg PC and 15 molar % dicetyl phosphate. Acceptor vesicles consisted of 80 molar % egg PC and 20 molar % cholesterol; 0.1 mg/ml of donor vesicles was mixed with 1 mg/ml of acceptor vesicles and separated on ion-exchange columns as described [21]. Donor vesicles contained either [4-^{14}C]cholesterol (◆) or 1-palmitoyl-2-oleoyl [1-^{14}C]PC (POPC; ●). The inset shows the fraction of [^{14}C]cholesterol and [^{14}C]POPC counts remaining in donor vesicles as a function of time. The transfer of both species is first order with respect to its concentration in the donor vesicles and the half-time for cholesterol exchange is 2.3 ± 0.3 h [21] with all of the cholesterol in the vesicles apparently being in a single kinetic pool. If transmembrane movement of POPC is assumed to be rapid relative to the rate of exchange, then the $t_{1/2}$ for exchange is 48 ± 5 h [21]. If only POPC in the outer monolayer of the bilayer is available for exchange, however, then the $t_{1/2}$ is 63 h [26].

Fig. 5. Cholesterol exchange from 1 molar % cholesterol/ PC donor vesicles of differing acyl chain compositions

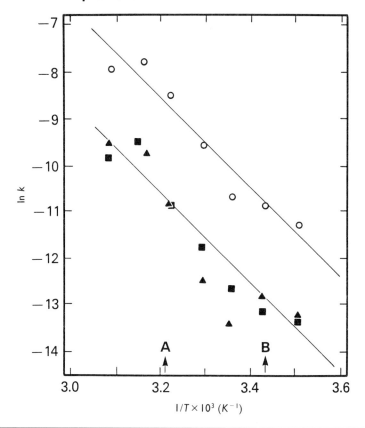

The rate of [4-^{14}C]cholesterol transfer from 1 molar % cholesterol and 99 molar % PC vesicles of differing acyl chain compositions to negatively charged vesicles of similar composition was followed over a range of temperature as described [25]. The host lipid consisted of egg yolk PC (○), dimyristoyl PC (■) or dipalmitoyl PC (▲). The lines are the best fit to the two groups of data by linear regression. The gel to liquid-crystal transition temperatures of dipalmitoyl PC and dimyristoyl PC unilamellar vesicles are shown by arrows A and B, respectively. Reproduced from [25] with permission.

demonstrate that this effect is observed because increased unsaturation decreases $t_{1/2}$. The increase in rate constant for desorption with increased unsaturation of the host phospholipid matrix has been observed in several laboratories (for a review, see [17]). As the cholesterol molecule is neutral it is unlikely that electrostatic interactions will influence the magnitude of ΔG^* for formation of the transition state in Fig. 6. In agreement with this,

Fig. 6. Free-energy diagram of lipid exchange from bilayers

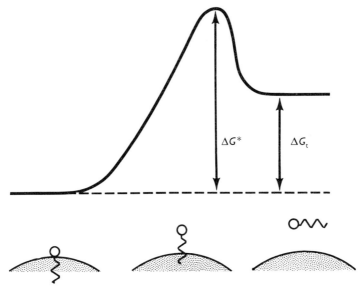

Transfer of lipid molecules from the bilayer to the aqueous phase proceeds through a transition-state complex, where the lipid molecule is attached to the vesicle by the tip of its hydrophobic tail. Formation of this activated state complex is associated with a high energy of activation, which exceeds the free energy of transfer by an amount that is determined, in part, by restriction of the lipid molecule to the surface of the bilayer vesicle. Reproduced from [26] with permission.

addition of up to 15 molar % dicetyl phosphate [23] to confer a negative charge on the donor egg PC/cholesterol vesicle does not significantly alter $t_{1/2}$ for cholesterol exchange. However, the type of phospholipid used to make the donor vesicle can modify $t_{1/2}$ for cholesterol exchange. Consistent with a minor role for surface charge, Bittman and associates [24] have shown that the $t_{1/2}$ for cholesterol desorption from vesicles of egg PC, egg phosphatidylethanolamine and phosphatidylglycerol are similar. The effect of incorporating sphingomyelin into the donor bilayer is in striking contrast because significant reductions in the rate of cholesterol exchange are observed (Table 1). This implies that the cholesterol–sphingomyelin intermolecular interaction energy is greater than that for cholesterol with the other phospholipids.

The differences in cholesterol–phospholipid interaction energy in the donor bilayer, which give rise to the above ranking of rate constants for cholesterol exchange, may be a consequence of variations in several factors. One such factor is the relative juxtaposition of cholesterol and phospholipid molecules along their long axes (i.e., normal to the bilayer plane). However, spectroscopic data indicate that the exposure of cholesterol molecules to the aqueous phase is similar in all the bilayers, with the cholesterol hydroxyl group located in the region of the phospholipid carbonyl groups. In the case of PC bilayers, the cholesterol molecule extends to about the C_{12} atom of the

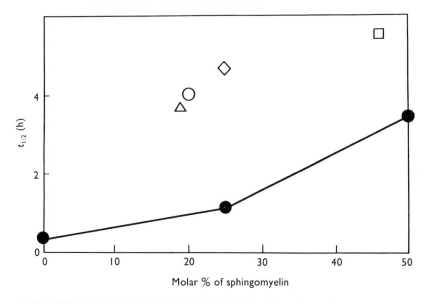

Fig. 7. The influence of the sphingomyelin content of red blood cells and egg PC unilamellar vesicles on the kinetics of cholesterol efflux

The molar % of sphingomyelin plotted on the abscissa is relative to the total phospholipid composition of the membrane. The half-time for transfer of [^{14}C]cholesterol from egg PC/egg sphingomyelin vesicles at 45 °C is illustrated (●). Also shown is the $t_{1/2}$ for efflux of cholesterol from different species of red blood cells at 37 °C: ○, rabbit; △, rat; ◇, human; □, cow. Reproduced from [28] with permission.

acyl chains (Fig. 2). This location means that the unsaturated chains in, for example, brain and egg sphingomyelin are effectively saturated as far as their interaction with cholesterol is concerned because the only unsaturated fatty acid is nervonic acid and the double bond is at the C_{15} atom. A second factor affecting cholesterol–phospholipid interaction energy is the lateral packing density of the cholesterol and phospholipid molecules in the bilayer. In relatively expanded bilayers, the Van der Waals attractive interaction energy between cholesterol and the phospholipid hydrocarbon chains will be lower. Molecular area data indicate that the relative expansions of the phospholipid molecules (in an equimolar mixture with cholesterol) are in the rank

Table I. Half-times for exchange of cholesterol from small unilamellar phospholipid vesicles

Host phospholipid*	$t_{1/2}$ (h)†	
	37 °C	45 °C
Phosphatidylcholine		
Dipalmitoyl	16 ± 3 (n = 10)	5.5 ± 0.8 (n = 20)
Dimyristoyl	11 ± 1 (n = 9)	
Egg	1.2 ± 0.05 (n = 19)	0.5 ± 0.04 (n = 6)
Dioleoyl	0.8 ± 0.09 (n = 11)	0.15 ± 0.03 (n = 4)
Soybean	0.3 ± 0.06 (n = 13)	
Sphingomyelin		
Egg		19 ± 1 (n = 10)
Bovine brain		14 ± 3 (n = 10)
Palmitoyl	77 ± 4 (n = 7)	11 ± 2 (n = 12)

*The donor vesicles consisted of 10 molar % [^{14}C]cholesterol, 15 molar % dicetyl phosphate and 75 molar % phospholipid. The donor vesicles were used in cholesterol-exchange experiments with 10-fold excess acceptor vesicles (10 molar % cholesterol, 90 molar % egg PC and a trace of [^{3}H]-cholesteryl oleate) present so that the $t_{1/2}$ values are characteristic of the rate-limiting desorption of cholesterol from the phospholipid–water interface (see [21]). †The half-times are derived from the first-order rate constants (k_1) describing the kinetics of cholesterol exchange using the relationship $t_{1/2} = (\ln 2/k_1)$. The tabulated $t_{1/2}$ values are mean values (± S.E.M.) and n is the number of kinetic experiments performed. At least three different preparations of vesicles were used except in the experiments with unsaturated PC at 45 °C. Reproduced from [22] with permission.

---

unsaturated PC > saturated PC > sphingomyelin [22]. This indicates that the Van der Waals attraction between cholesterol and phospholipid increases in the same order.

If the simplifying assumption is made that the differences in the free energy of activation, which underlie the ranking of rate constants for cholesterol exchange from bilayers in the order unsaturated PC > saturated PC > sphingomyelin, arise solely from variations in the cholesterol–phospholipid Van der Waals interaction energy, then more quantitative insights can be gained. Using an approximate expression for the calculation of the Van der Waals attractive interaction energy between cholesterol and phospholipid molecules [22], the ratio of the rate constants (k_1 and k_2) for cholesterol desorption from bilayers of two different phospholipids (designated by the subscripts 1 and 2) is given by:

$$\ln(k_1/k_2) = \frac{10^4}{RT}\left[\frac{1}{A_2^{5/2}} - \frac{1}{A_1^{5/2}}\right] \quad (1)$$

where A (in Å^2) is half the molecular area occupied by a phospholipid or cholesterol molecule, R is the gas constant and T is the absolute temperature (the product is expressed in kcal/mol). Eqn. (1), which gives the ratio of the rate constants for cholesterol desorption as a function of the relative molecular areas in two bilayers, has been applied to the $t_{1/2}$ values listed in Table 1 for cholesterol exchange from vesicles of unsaturated PC, saturated PC and sphingomyelin [22]. The results of the analysis are consistent with differences in cholesterol–phospholipid Van der Waals interaction energy being the cause of varying rates of cholesterol exchange from different host phospholipid bilayers.

Since cholesterol condenses a liquid phospholipid monolayer, thereby raising the lateral packing density and decreasing the average molecular area, from eqn. (1) the $t_{1/2}$ of desorption might be expected to increase as the cholesterol/phospholipid ratio is increased. However, $t_{1/2}$ is not strongly dependent on this ratio [25]; this has been attributed to the occurrence of lateral phase separation in phospholipid/cholesterol monolayers and bilayers so that the stoichiometry and local environment of any given cholesterol molecule does not vary greatly as the overall cholesterol content rises (see previous section and Fig. 3). Consistent with the occurrence of clusters of PC/cholesterol molecules separated laterally from free PC and unable to undergo the gel to liquid-crystal phase transition, the experimental activation energy for cholesterol desorption is insensitive either to the cholesterol:PC ratio in egg PC vesicles or to the occurrence of the gel to liquid-crystal phase transition in dimyristoyl PC and dipalmitoyl PC bilayers (Fig. 5). In marked contrast, the rates of desorption of dimyristoyl PC [26] and sphingomyelin [27] molecules are markedly reduced as the bilayer is cooled below the chain-melting temperature because the transferring phospholipid molecule has to desorb from the tightly packed gel phase.

Comparison of cell membranes and model membranes

The above studies with small unilamellar vesicles of defined phospholipid composition show that increases in the lateral packing density in the lipid–water interface decrease the rate of cholesterol desorption because the higher cholesterol–phospholipid Van der Waals attraction raises the activation energy barrier opposing desorption (Fig. 6). Increased Van der Waals attraction between sphingomyelin and cholesterol relative to PC and cholesterol is also significant in explaining the slower exchange from the sphingomyelin system. It is important to establish whether or not the above effects observed in model membranes also occur in the plasma membranes of cells.

Fig. 7 depicts data which address this point. The $t_{1/2}$ for efflux of cholesterol from four species of red blood cells and from egg PC/egg sphingomyelin vesicles are plotted as a function of the molar % sphingomyelin in the donor membrane. Qualitatively, $t_{1/2}$ increases with increasing sphingomyelin content in both systems. However, the influence of increasing the sphingomyelin level is greater in the egg PC/egg sphingomyelin vesicle system. For example, raising sphingomyelin from about 20 to 50% is associated with a threefold increase in $t_{1/2}$ from about 1–3 h, whereas, in red blood cells $t_{1/2}$ increases by a factor of only about 1.5 (3.7 h for rat red blood cells to 5.5 h for bovine red blood cells). Factors that contribute to the quantitative differences between small unilamellar vesicles and red blood cells probably include the differences in cholesterol content (10 molar % in vesicles and about 40 molar % in red blood cells) and the complex acyl chain composition of red blood cells compared with that of the egg PC/egg sphingomyelin mixed bilayer. In addition, the composition of the outer leaflet of the membrane is presumed to have the major effect on the rate of cholesterol desorption into the extracellular aqueous phase. The molar % sphingomyelin plotted on the abscissa of Fig. 7 reflects the total phospholipid composition of the membrane and would only reflect the composition of the outer monolayer if the transbilayer phospholipid distribution were symmetrical. In fact, sphingomyelin is located preferentially in the outer monolayer (80% of the sphingomyelin is in this location in human red blood cells [29]), so that the sphingomyelin content of the outer monolayer of red blood cells is under-represented on the abscissa of Fig. 7.

Overall, the data summarized in Fig. 7 suggests that increasing the membrane sphingomyelin:PC ratio raises the $t_{1/2}$ for cholesterol exchange in both the red blood cell plasma membrane and in simple mixed sphingomyelin PC bilayers. However, the influence of sphingomyelin is less than the natural plasma membrane, perhaps because of modulating factors such as membrane proteins and the presence of a complex phospholipid mixture. More work is required to elucidate the roles of the various plasma membrane components in cholesterol efflux.

Conclusions

The effects of cholesterol on the packing of phospholipid acyl chains and bilayer membranes are now understood in detail. This development has been made on the basis of pioneering studies, conducted by Dennis Chapman and colleagues on phospholipid phase behaviour and the modulation of this behaviour by cholesterol. The dynamics of cholesterol molecules within a membrane are important for the function of the membrane. As reviewed here, cholesterol molecules can also diffuse between membranes in a biologically significant time scale and the rate is extremely sensitive to the cholesterol–phospholipid interactions in the donor membrane. Future work should provide insights into the effects of membrane domains and membrane proteins on both the intracellular trafficking of cholesterol and the transport of plasma membrane cholesterol to extracellular lipoprotein acceptor particles.

The work from this laboratory was supported by National Institutes of Health Grants HL22633 and HL07443.

References

1. Chapman, D. & Penkett, S. A. (1966) Nature (London) 211, 1304–1305
2. Oldfield, E. & Chapman, D. (1972) FEBS Lett. 23, 285–297
3. Quinn, P. J. & Chapman, D. (1980) CRC Crit. Rev. Biochem. 8, 1–117
4. Chapman, D. & Hayward, J. A. (1985) Biochem. J. 228, 281–295
5. Lee, D. C. & Chapman, D. (1986) Biosci. Rep. 6, 235–256
6. Darke, A., Finer, E. G., Flook, A. G. & Phillips, M. C. (1972) J. Mol. Biol. 63, 265–279
7. Bourges, M., Small, D. M. & Dervichian, D. G. (1967) Biochim. Biophys. Acta 137, 157–167
8. Collins, J. J. & Phillips, M. C. (1982) J. Lipid Res. 23, 291–298
9. Phillips, M. C. (1972) Prog. Surf. Membr. Sci. 5, 139–221
10. Yeagle, P. L. (1988) Biology of Cholesterol, CRC Press, Boca Raton
11. Shimshick, E. J. & McConnell, H. M. (1973) Biochem. Biophys. Res. Commun. 53, 446–451
12. Sankaram, M. B. & Thompson, T. E. (1990) Biochemistry 29, 10670–10675
13. Hui, S. W. (1988) in Biology of Cholesterol (P. L. Yeagle, ed.), pp. 213–231, CRC Press, Boca Raton
14. Ipsen, J. H., Karlstrom, G., Mouritsen, O. G., Wennerstrom, H. & Zuckerman, M. J. (1987) Biochim. Biophys. Acta 905, 162–172
15. Sankaram, M. B. & Thompson, T. E. (1990) Biochemistry 29, 10676–10684
16. Presti, F. T., Pace, R. J. & Chan, S. I. (1982) Biochemistry 21, 3831–3835
17. Phillips, M. C., Johnson, W. J. & Rothblat, G. H. (1987) Biochim. Biophys. Acta 906, 223–276
18. Johnson, W. J., Mahlberg, F. H., Rothblat, G. H. & Phillips, M. C. (1991) Biochim. Biophys. Acta 1085, 273–298
19. Dawidowicz, E. A. (1987) Curr. Top. Membr. Trans. 29, 175–202
20. Rothblat, G. H. & Phillips, M. C. (1991) Curr. Opin. Lipidol. 2, 288–294
21. McLean, L. R. & Phillips, M. C. (1981) Biochemistry 20, 2893–2900
22. Lund-Katz, S., Laboda, H. M., McLean, L. R. & Phillips, M. C. (1988) Biochemistry 27, 3416–3423
23. McLean, L. R. & Phillips, M. C. (1984) Biochim. Biophys. Acta 776, 21–26
24. Fugler, L., Clejan, S. & Bittman, R. (1985) J. Biol. Chem. 260, 4098–4102
25. McLean, L. R. & Phillips, M. C. (1982) Biochemistry 21, 4053–4059
26. McLean, L. R. & Phillips, M. C. (1984) Biochemistry 23, 4624–4630
27. Frank, A., Barenholz, Y., Lichtenberg, D. & Thompson, T. E. (1983) Biochemistry 22, 5647–5651
28. Gold, J. C. & Phillips, M. C. (1990) Biochim. Biophys. Acta 1027, 85–92
29. Verkleij, A. J., Zwaal, R. F. A., Roelofsen, B., Comfurius, P., Kastelijn, D. & Van Deenen, L. L. M. (1973) Biochim. Biophys. Acta 323, 178–193

Developments in the nuclear magnetic resonance spectroscopy of lipids and membranes: the first quarter century

Eric Oldfield

Department of Chemistry, University of Illinois at Urbana-Champaign, 505 South Mathews Avenue, Urbana, IL 61801, U.S.A.

Introduction

In 1966, Chapman and Penkett [1] reported the first observation of high-resolution n.m.r. spectra of model membranes, and demonstrated in a convincing way the condensing effect of cholesterol on phospholipid acyl chains, at the molecular level. The lecithin/cholesterol system continues to be a paradigm for real biological membranes, such as the red cell and myelin, and over the past 25 years just about every new n.m.r. technique has been applied towards gaining an understanding of this complex system. In the following, we give a brief account of the progress in this area, right up until the present, where extremely highly resolved spectra of both model and biological membranes can now be obtained, which include some observations on membrane proteins themselves.

Developments in ^{1}H n.m.r. spectroscopy

We began our studies of lipids in membranes in 1968, using sonicated single-bilayer systems. However, there were always reservations about using sonication, especially for biomembranes, and thus new approaches had to be developed. One expansion of the very early ^{1}H n.m.r. experiments suggested by Chapman involved the use of the so-called 'magic-angle' sample spinning (MAS) method, in which lipids are spun rapidly (several kHz) in a gas turbine and the dipolar interactions between protons, which broaden static n.m.r. spectra, are averaged out as a result of sample rotation. Working with

Doskočilová and Schneider, we obtained modest resolution improvements [2], but the time was not yet right, as sensitivity and resolution at the low magnetic field strengths in use in 1972 (~1.4 Tesla) were simply inadequate for true high-resolution, although the results did look promising. Some 15 years later, we tried the same experiments again, this time using a high-field magnet (11.7 Tesla, 500 MHz for protons), together with the Fourier transform technique, and obtained very well resolved spectra, as shown in Fig. 1. The results we obtained were quite surprising in that the extremely broad 'super-Lorentzian' lineshapes observed by Lawson and Flautt [3] for smectic liquid crystals, which obscure high resolution features, broke up at even low spinning speeds into a manifold of sharp spinning sidebands [4], and truly high-resolution spectra were obtained [4, 5].

The observation of such high-resolution ^{1}H MAS n.m.r. spectra was surprising, since the static dipolar Hamiltonian:

$$\sum_{i<j} D_{ij}(\phi)(3I_{zi}I_{zj} - I_i \cdot I_j) \tag{1}$$

does not in general commute with itself at different rotor orientations, ϕ. However, in most liquid crystalline phases, intermolecular dipole–dipole interactions are averaged by fast lateral diffusion, while fast axial rotation reduces the intramolecular dipole–dipole interaction and causes the angular dependence of the Hamiltonian to be the same for all proton pairs. Thus, the dipolar interaction is scaled by $P_2 (\cos\theta)$, where θ is the angle between the director axis and H_o, such that

$$D_{ij} = \tfrac{1}{2}(3\cos^2\theta - 1)D_{ij}^o \tag{2}$$

and the Hamiltonian becomes

$$\tfrac{1}{2}(3\cos^2\theta - 1)\sum_{i<j} D_{ij}^o(3I_{zi}I_{zj} - I_i \cdot I_j) \tag{3}$$

which under MAS commutes with itself at different rotor orientations, ϕ. The result is that the dipole–dipole interaction becomes inhomogeneous and sharp spinning sidebands are obtained at sample rotation rates much slower than the static width. In addition, we also found that the envelope of the spinning sidebands in a slow-spin spectrum was very close to the theoretical 'super-Lorentzian' lineshape [3] given by [6]:

$$L(v-v_o) = \int_0^1 |3\cos^2\theta - 1|^{-1} \times f[(v-v_o)/|3\cos^2\theta - 1|]\, d\cos\theta \tag{4}$$

Similar breakdown of the static lineshape into numerous sharp spinning sidebands has now been observed for most of the other smectic liquid crystalline phase lipids we have investigated, including monogalactosyldiglyceride, digalactosyldiglyceride, sulphoquinovosyldiglyceride, egg

Fig. 1. 500 MHz ^{1}H 'magic-angle' sample spinning n.m.r. spectrum of dimyristoylphosphatidylcholine/H$_2$O (50 wt.% ^{2}H$_2$O) at 30 °C

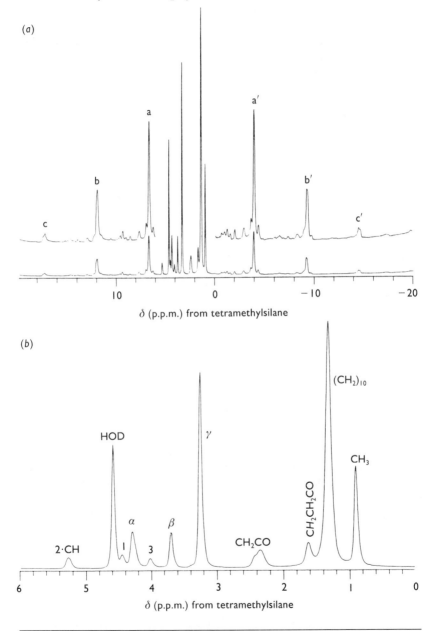

(a) Full spectrum; (b) centre-band region only. Reprinted from [5] with permission.

phosphatidylcholine, beef brain sphingomyelin, beef heart cardiolipin, potassium oleate-water, phosphatidylinositol, dioleoylphosphatidylcholine and dipalmitoylphosphatidylcholine.

We also found that molecular order parameters could, in some instances, be deduced from the side-band patterns [5], the results obtained agreeing strongly with those found using ^{2}H n.m.r. More recently, we have investigated the proton spin-lattice relaxation in lipid bilayers, but found only a rather small graduation in relaxation rates from one part of the lipid to another. This re-ignited the old controversy about the occurrence of spin-diffusion — the transfer of magnetization from one part of the lipid to another by means of energy-conserving mutual spin flip-flops, and the results of two dimensional n.m.r. experiments using isotopically ^{2}H-labelled molecules have tended towards a view that spin-diffusion does occur in these systems. This topic is now of particular interest in the area of contrast enhancement in magnetic resonance imaging (MRI).

Using MRI to study brain, it has been shown that white matter in myelin, which contains very large amounts of cholesterol, appears extremely 'bright' in relaxation-weighted MRI images, which means that the H_2O protons are relaxed much more effectively than in grey matter. These MRI images have been used to investigate brain development in infants, as well as the breakdown of myelin in e.g. multiple sclerosis and Koenig, Balabun and their colleagues [7, 8] have shown that cholesterol is likely to make a major contribution to the relaxation enhancement (or brightness). Looking at egg lecthin with and without cholesterol (the basic Chapman and Penkett system of 1966), these authors have found that water relaxation increases monotonically with cholesterol content and then flattens out at a ratio of 1:1 lecithin/cholesterol, as might be expected on the basis of known phase diagrams [9]. The results of other experiments have shown that the H_2O signal intensity can be altered by means of off-resonance irradiation of the lecithin/sterol protons, indicating cross-relaxation or spin-diffusion between H_2O and lipid, the large relaxation enhancement apparently being due to the presence of cholesterol (as demonstrated by use of a deuterated lipid plus cholesterol, where the same enhancement is seen). These experiments appear to fit in nicely with our own [4, 5] in which we have found that the cholesterol ring's protons are not visible in ^{1}H MAS experiments probably because of strong dipolar interactions in the cholesterol spin system, which is sufficient to bring about effective spin-diffusion.

In our laboratory, we have found that other interactions may also be important in H_2O relaxation. For example, as shown in Table 1, we find extremely rapid H_2O relaxation in the presence of the glycolipid digalactosyldiglyceride (DGDG), presumably because of strong interactions of H_2O with the sugar hydroxyl groups.

Because around one-third of the dry weight of myelin lipid consists of the glycosphingolipid, cerebroside, it seems likely that cerebroside galactosyl–water interactions could also play a role in contrast enhancement.

Table 1. Water proton spin-lattice relaxation times in lipid/water systems (K. D. Park & E. Oldfield, unpublished work)

System	T_1 H$_2$O (s)
Egg lecithin	2.2
Egg lecithin/cholesterol (1:1)	1.6
Digalactosyldiglyceride	1.1

Lipids are 1:1 wt ratio with H$_2$O.

Thus, the early work on lecithin/cholesterol dynamics and phase behaviour has gradually evolved, through studies of two-dimensional n.m.r. spin-diffusion and cross-relaxation, to the rapidly developing area of MRI contrast enhancement, where cholesterol has been shown to play a major role [7, 8] in differentiating grey from white matter, in both development and disease.

Applications of ^{2}H n.m.r. spectroscopy

The cholesterol story took a major turn in 1971 when, in collaboration with Chapman and Derbyshire (at Nottingham), we were able to show that ^{2}H n.m.r. spectroscopy of ^{2}H-labelled lipids was likely to be a rather useful technique for studying lipid (and later, protein) structure. Our initial studies involved the massive ^{2}H-labelling of the acyl side chains of 1,2-dimyristoyl-sn-glycero-3-phosphocholine (dimyristoylphosphatidylcholine, DMPC), and wide-line, continuous-wave n.m.r. [10], and we were pleased to see that gel phase, liquid-crystal phase and DMPC-cholesterol/water systems all had quite different spectra. The major quadrupole splitting (the 'plateau') of DMPC in the liquid-crystal phase was ~27 kHz, and this increased to ~50 kHz upon addition of an equimolar amount of cholesterol, which translates to about a factor of two increase in molecular order parameter. The ~50 kHz splitting remained below the pure lipid T_c, reflecting the 'state of intermediate fluidity' or the presence of the now so-called liquid-ordered phase (lo). The ^{2}H n.m.r. technique was then developed rapidly by Seelig in Basel [11] using specifically deuterated lipids, and its progression still continues unabated (see [12-14] for reviews).

Two further useful developments in the ^{2}H n.m.r. area came from the works of Bloom, Davis and their colleagues [15, 16]. The first involved use of a spin-echo method [15], which permits the acquisition of relatively undistorted ^{2}H n.m.r. spectra, which would otherwise be distorted in a one-pulse technique as a consequence of pulse-feedthrough. The second involved numerical methods to remove the broadening caused by the powder pattern distribution — the 'Pake doublet', dubbed 'de-Paking' by Bloom and colleagues ([16] and references therein). We have now added a second deconvolution method, the Levenburg–Marquart algorithm [17], in which de-Paked spectra are further deconvoluted to give individual resonances for each labelled site, and a typical result on a 1-[^{2}H$_{31}$]-palmitoyl-2[(9 Z)octadec-9-enoyl]-sn-glycero-3-phosphocholine (palmitoyl oleoyl phophatidylcholine, POPC/cholesterol system (3:1 molar ratio, 1:1 wt ratio with H$_2$O) is shown in Fig. 2. The resolution is very good and we believe the method permits more accurate spectral analyses.

Using such well resolved spectra, we have investigated the condensing effects of a number of sterols other than cholesterol (H. Le & E. Oldfield, unpublished work) and have found that cholesterol is the most effective sterol in causing chain condensation. Addition of e.g. ring methyl groups causes decreased ordering compared with the effect of cholesterol, and could be explained by a tendency towards a surface 'ruffling' effect of the sterol nucleus. The results of an extensive series of studies in the late 1970s and early 1980s by our group, Bloom, Dahlquist, Seelig and others [18–22], led to a general picture of protein–lipid interaction in membranes in which there was little ordering of lipid acyl chains by protein. Rather, chain (and headgroup [22]) motions appear to slow down considerably in the presence of protein but, in general, the chains do not order in the way they do with cholesterol.

The ^{2}H n.m.r. method is not restricted to studies of hydrocarbon chain order and dynamics, as shown in early work by Seelig [23]. Unfortunately, it is not yet possible to unambiguously assign headgroup conformations for very flexible headgroups, like choline, on the basis of n.m.r. results alone. Much better progress has been made with glycolipids such as cerebroside since there are a number of fixed connectivities in the sugar group. With multiple-site labelling, good results for headgroup orientation can be obtained, both for cerebrosides [24] and many mono- and diglycosyl-diglycerides [25]. A representative picture of cerebroside orientations in model membranes is given in Fig. 3, and analogous results have been obtained for the glycosylglycerolipids in elegant ^{2}H n.m.r. studies performed by Smith, Jarrell and their colleagues [25]. It seems plausible that this basic extended conformation applies to the galactosyl headgroups of cerebrosides in myelin, although this will be difficult to test experimentally.

^{13}C n.m.r. studies

So far, I have discussed results obtained for lipid, lipid/sterol and lipid/protein systems. The next step is, of course, to investigate what happens when all three components are mixed together and to try to see how the interactions in the model system compare with those seen in real biological membranes. Proton n.m.r. spectra of even the lipid/sterol systems are of limited use because the sterol protons are, for the most part, 'invisible' under high-resolution MAS n.m.r. conditions. Deuterium n.m.r. is a good candidate — at least for model systems — and we have found in model DMPC/

Fig. 2. 55.7 MHz (8.45 Tesla) ^{2}H n.m.r. spectra of POPC/cholesterol (3:1 molar ratio, 50 wt% H_2O) at 37 °C

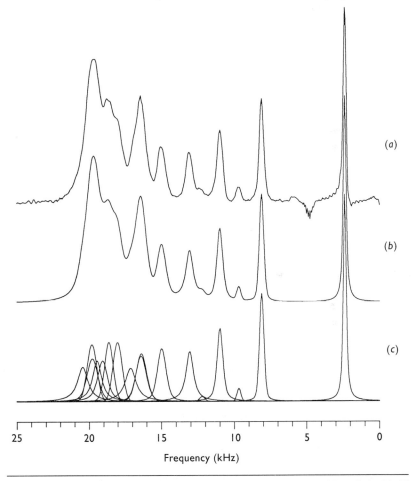

(a) De-Paked half spectrum; (b) deconvoluted (Levenburg–Marquardt) de-Paked half spectrum; (c) individual component lines in (b).

cholesterol/gramicidin systems, using ^{2}H-labelled lipids, that cholesterol dominates the ^{2}H quadrupole splitting (J. Patterson, T. Bowers & E. Oldfield, unpublished work), although the rates of motion appear to decrease. ^{2}H n.m.r. appears to be very useful for investigating the model-

Fig. 3. Cerebroside headgroup orientations deduced from a conformational analysis of ^{2}H n.m.r. spectra of selectively deuterated *N*-hexadecanoylglucocerebroside (PGLC), at 90 °C

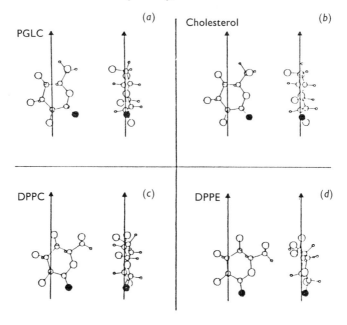

(a) PGLC; (b) 1:1 molar ratio PGLC/cholesterol; (c) 17 wt% PGLC in DPPC; (d) 17 wt% PGLC in dipalmitoylphosphatidylethanolamine. The arrows represent the bilayer normal, and the solid circles are the glycosidic oxygens. The structure on the right of each pair is generated by a 90° rotation of the left-hand structure about the bilayer normal. The hydroxymethyl group is shown to aid visualization of the ring, but the A_6 torsion angle is arbitrary. Reprinted from [24] with permission.

compound-containing systems, but is going to be of limited use in studies of many biomembranes, e.g. myelin and the red cell. However, as ourselves and others showed early on [26, 27] the technique can be used to investigate the structures of some biological membranes, i.e. those that can be biosynthetically enriched with ^{2}H-labelled fatty acids (or sterols).

What is required for more general studies of 'intact' biomembrane structure is a high resolution, natural abundance technique, and ^{13}C n.m.r. fits the bill.

Our earliest ^{13}C n.m.r. attempts [28, 29] on unsonicated multi-bilayers with and without cholesterol and on rat liver mitochondrial membranes, although of limited resolution and sensitivity, were nonetheless encouraging, and we asserted that 'studies on ^{13}C nuclei in higher fields, with lipids and membranes, where resolution is enhanced by an increased chemical shift range may be particularly useful' [2]. We were correct, but it took over a decade of developments in magnet technology to get to the point where these resolution (and sensitivity) improvements were actually achievable.

We show in Fig. 4 typical ^{13}C MAS n.m.r. results on lecithin and lecithin/cholesterol bilayers, together with conventional 'solution' ^{13}C n.m.r. results, without MAS, on sonicated bilayers [30]. For reasons that are well understood, the MAS n.m.r. technique gives remarkably better resolved ^{13}C spectra of the rigid sterol nucleus, which has led to the application of the technique to an analysis of an intact cell membrane, in this case the myelin membrane, from human brain [5]. Fig. 5 shows the 11.7 Tesla (500 MHz ^{1}H, 125 MHz ^{13}C) ^{13}C MAS n.m.r. spectrum of a sample of myelin taken from a neurologically normal adult, in which a large number of resonances can be resolved and assigned [5]. Such resolution is by far the best achieved to date on an intact biological membrane, the myelin membrane being a particularly attractive one for study because of its low water content, high lipid and sterol content, and the possible implications of any results obtained to human disease, an example being the contrast enhancement observed in MRI of abnormal and developing human myelins.

In ^{13}C n.m.r., the direct order-parameter information seen in ^{2}H n.m.r. is lost, but the trade-off is that many new systems which cannot be labelled, e.g. atherosclerotic plaque, lung surfactant, red cell membranes and myelin, become amenable to study. In addition, with ^{13}C n.m.r., there are qualitatively different types of spectroscopic parameters which can be used as structural probes, examples being the chemical shift [5], radiofrequency field-induced transverse relaxation (F. Adebodun, J. Chung, B. Montez, E. Oldfield & X. Shan, unpublished work), differential linebroadening [31, 32] and cross-relaxation. Figs. 6–9 give some idea of what might be in store. For example, we show in Fig. 6 ^{13}C MAS n.m.r. spectra of the natural plant glycolipid digalactosyldiglyceride (as a 1:1 wt ratio dispersion in H_2O) at 20, -10 and -30 °C. At 20 °C, essentially all carbons in the molecule are very well resolved, while at -30 °C, all of the headgroup, backbone carbons and the first two or three carbons of the acyl chain are broadened beyond detection. This can be explained by an interference effect between the coherent radiofrequency decoupling field and the incoherent thermal motions in the molecule and, as has been shown previously by others [33, 34], when the rate of motion of the C—H vector of interest is about the same as the radiofrequency field strength (in Hz), then decoupling is ineffective and considerable linebroadening ensues [32]. In DGDG, we see that the headgroup region is the first to become immobile (possibly coincidentally with bulk water freezing) on cooling, and that this effect is then transmitted along the acyl chains into the hydrocarbon interior. In our DGDG sample, 18:3 is the major fatty acid and, on cooling, it is possible to see, sequentially, Δ^9, Δ^{12} and then Δ^{15} 'freeze-out' or broaden. Such r.f. field-induced transverse relaxation could be a useful probe of lipid bilayer dynamics.

Fig. 4. 125 MHz (11.7 Tesla) proton-decoupled ^{13}C Fourier transform n.m.r. spectra at 21 °C of lecithin and lecithin/cholesterol (1:1) samples

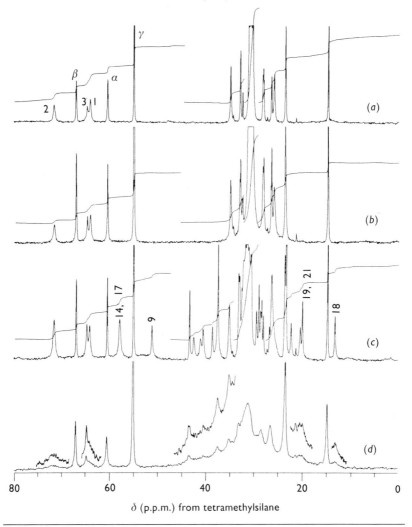

(a) Proton-decoupled ^{13}C MASS (2.1 kHz) n.m.r. spectrum of egg yolk lecithin (50 wt%)–D$_2$O at ~ 40-W continuous wave (CW) ^{1}H decoupling; (b) nuclear Overhauser effect enhanced, proton-decoupled ^{13}C n.m.r. spectrum of sonicated egg yolk lecithin (12.7 wt%)–D$_2$O at ~ 5-W WALTZ-16 ^{1}H decoupling; (c) proton-decoupled ^{13}C MASS (2.8 kHz) n.m.r. spectrum of egg yolk lecithin (33 wt%) cholesterol (17 wt%)–D$_2$O (50 wt%) at ~ 40-W CW ^{1}H decoupling; (d) nuclear Overhauser effect enhanced, proton-decoupled ^{13}C n.m.r. spectrum of sonicated egg yolk lecithin (8.5 wt%)–cholesterol (4.2 wt%)–D$_2$O (87.3 wt%) at ~ 5-W WALTZ-16 ^{1}H decoupling. An exponential line broadening of 6 Hz was applied to each spectrum. Recycle times of 5 s were used in all spectra shown. Reproduced from [30] with permission.

Fig. 5. 125.7 MHz (11.7 Tesla) proton-decoupled ^{13}C Fourier transform MAS n.m.r. spectrum of human adult myelin (43 years, male, myocardial infarct) at 37 °C

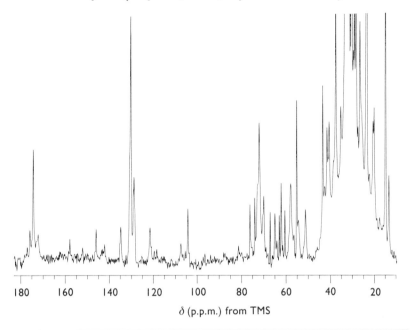

Spectral conditions are as follows: 3 kHz MAS rate, 16452 accumulations, 5 s recycle time, 2X 8k data points, ^{13}C 90° pulse width = 9 µs, 10 Hz linebroadening due to exponential multiplication.

A second interesting phenomenon that we have observed in liquid-crystalline phase bilayers, as well as in nematic and cholesteric liquid crystals and some polymers, is that of differential linebroadening of 'J-coupled' multiplets ([32] and references therein). An excellent example of this effect is shown in Fig. 7, where we are observing the proton-coupled ^{13}C MAS n.m.r. spectrum of the tricyclic antidepressant drug, desipramine, in a DMPC bilayer — a system previously investigated by Chapman, Cater and colleagues [35]. The differential linebroadening effect for the desipramine sp^2 carbons is thought to be caused by interference effects between the ^{13}C—^{1}H dipolar and the ^{13}C chemical shift anisotropy interactions, the result of which is that each component of the C—H doublet has a different relaxation rate, or linewidth. However, the effects are not equal for each of the four ring methine carbons: two have large differential linebroadening effects and two have much smaller ones, as may be seen in Fig. 7. The likely origin of these different linewidths is the motional averaging of the C—H dipolar interaction: two of the ring C—H vectors are approximately along the $C_{(2)}$

Fig. 6. 125.7 MHz (11.7 Tesla) proton-decoupled ^{13}C MAS n.m.r. spectra of a natural plant digalactosyldiglyceride, at the temperatures indicated

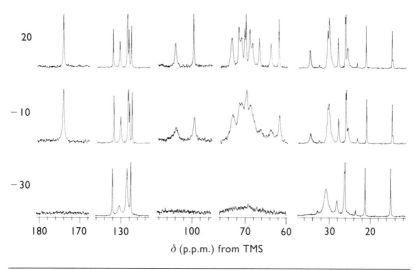

Spectral conditions are basically as described in Fig. 5.

axis of the molecule and two are close to the 'magic-angle' of 54.7°. Thus, fast rotation about the ring $C_{(2)}$ axis causes averaging of the dipolar interaction for C2 and C3 whereas the dipolar interactions at C1 and C4 are not appreciably averaged, and these sites have a large differential linebroadening. Similar effects are seen in other parts of the DMPC or DMPC/desipramine bilayer system and, certainly for model systems, coupled MAS n.m.r. spectra are likely to be a source of useful new information.

The third effect we wish to mention in our brief discussion of ^{13}C n.m.r. is that of cross-polarization (CP) [36]. The general technique of cross-polarization 'magic-angle' sample-spinning (CP-MAS) is by now probably well known, although we have not generally employed it in our membrane studies, as accurate estimates of peak intensities are difficult to achieve. The reason for this, however, is very interesting.

Now, in rigid solids, the rate at which ^{13}C magnetization increases in the CP experiment, for many protonated carbons, is fairly constant and of the order of several hundred microseconds. Membranes, on the other hand, have a wide range of rates and types of motion, and the rates at which magnetization builds up (and decays) vary over one or two orders of magnitude, as a result of the large range of motionally averaged dipolar interactions. A very typical result is seen in the two spectra shown in Fig. 8. In Fig. 8(a) we show a typical proton-decoupled Bloch decay spectrum of a

Fig. 7. 125.7 MHz (11.7 Tesla) proton-coupled ^{13}C MAS n.m.r. spectra of desipramine-1,2-ditetradecanoyl-sn-glycero-3-phosphocholine, 50 wt% water dispersion, at 40 °C

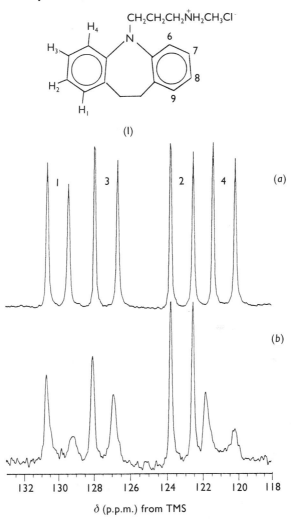

(a) 68.8 molar % desipramine (lipid basis); (b) 35.9 molar % desipramine; structure of desipramine shown above. Spectral conditions are basically as described in Fig. 5. The two broader doublets arise from C—H vectors (C1—^{1}H, C4—^{1}H) aligned approximately along the $C_{(2)}$ axis of desipramine, while the narrower doublets are thought to arise from C—H vectors aligned closer to the 'magic-angle' (C2—^{1}H, C3—^{1}H), result mix-time. With spectral editing, all mobile headgroup carbons, galactose C1–6, duced from [32] with permission.

Fig. 8. 125.7 MHz (11.7 Tesla) ^{13}C MAS n.m.r. spectra of 1,2-ditetradecanoyl-*sn*-glycero-3-phosphocholine, 50 wt% H$_2$O, at 37 °C

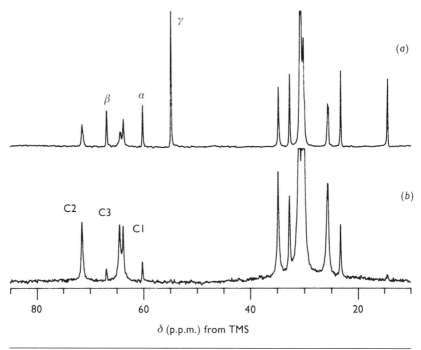

(a) Proton-decoupled Bloch decay spectrum showing approximately equal intensities for all types of carbon; (b) CP spectrum obtained using a 300 μs mix-time, showing spectral selection of rigid backbone and acyl side-chain carbons.

sample of DMPC, in excess water (50 wt% H$_2$O), at 37 °C. All peaks have close to the theoretical intensity on the basis of structural considerations. However, in a CP spectrum at short mix times, which accentuates the more rigid, dipolar-coupled spin signal intensities, a very different result is obtained (Fig. 8*b*) in which there is strong spectral editing in favour of the more rigid sites, there being little intensity from e.g. the mobile choline or terminal methyl groups. Thus, spectra can be edited on the basis of their 'solidity', or T_{CH}^{-1} cross-relaxation rate constants.

We believe this spectral-editing technique may have considerable use in simplifying complex biomembrane spectra. By way of example, we show in Fig. 9 the conventional ^{1}H-decoupled Bloch decay spectrum of a sample of human myelin (Fig. 9*a*) together with, for comparison, a spectrally edited myelin spectrum obtained using a 30 ms CP mix-time (Fig. 9*b*) which selects for the mobile groups (as opposed to the solid selection shown in Fig. 8*b*). Clearly, a remarkable simplification results from editing and, in the edited

Fig. 9. 125.7 MHz (11.7 Tesla) ^{13}C MAS n.m.r. spectra of human adult myelin, showing spectral selection of mobile headgroup carbon atom sites

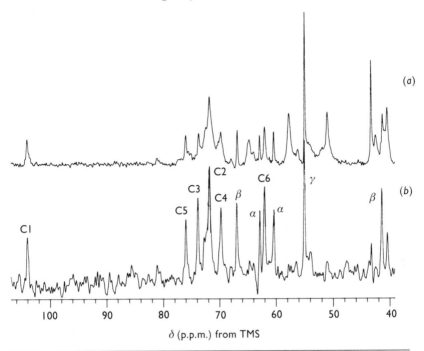

(a) Proton-decoupled Bloch decay spectrum; (b) CP spectrum obtained using a 30 ms mix-time. With spectral editing, all mobile headgroup carbons, galactose C1–6, PC/SM C^α, C^β, C^γ, and PE/PE-plasmalogen $C^{\alpha,\beta}$, can be resolved, demonstrating the high mobility of a major fraction of the myelin membrane headgroup sugar and phospholipid sites.

---

spectrum, all of the expected headgroup carbons, C1–6 of cerebroside, C^α, C^β and C^γ of phosphatidylcholine (PC)/sphingomyelin (SM) and C^α and C^β of phosphatidylethanolamine (PE) can be assigned — including the otherwise very difficult to resolve PE C^β and galactose $C_{(2,4)}$ carbons. The fact that many of the galactose, PC/SM and PE carbons are quite intense at long mix times indicates clearly that many of the myelin headgroups are highly mobile, the rigid carbons being edited on the basis of their short T_{CH} and ^{1}H-rotating frame relaxation times.

Conclusions

We have attempted to give a brief overview of the developments which have occurred over the past 25 years in the study of membrane structure by n.m.r.

spectroscopy — the original experiments being carried out in the mid to late 1960s by Dennis Chapman and his colleagues.

Perhaps not surprisingly, many of the systems currently under investigation remain the same as those of 25 years ago, examples being lecithin, lecithin/cholesterol and myelin. Lecithin was chosen in the early days because it gave at least one sharp peak, from the choline NMe_3 group, and myelin was chosen because of its high lipid content. Fortunately, things have now progressed to the point where, by using either ^2H spin-echo/de-Paking/deconvolution, or ^{13}C MAS n.m.r. (coupled MAS, CP MAS, decoupled MAS), essentially every atomic site in lecithin, lecithin/cholesterol — or any other model lipid system and, to some extent, intact biomembranes — can be probed by means of solid-state n.m.r. spectroscopy. A much better understanding of protein–lipid interaction has been obtained, and now a number of researchers are using ^2H and ^{13}C n.m.r. to investigate peptides and membrane proteins themselves [37–41]. The general topics of lipid–sterol (and possibly lipid–sterol–protein) interaction are even beginning to have some impact in areas such as brain (myelin) development and disease, as a consequence of the interactions that occur between brain H_2O (observed in MRI) and the membrane lipids. It will be interesting to see what happens in the next 25 years.

This work was supported by the United States Public Health Service (grant GM-40426).

References

1. Chapman, D. & Penkett, S. A. (1966) Nature (London) 211, 1304–1305
2. Chapman, D., Oldfield, E., Doskočilová, D. & Schneider, B. (1972) FEBS Lett. 25, 261–264
3. Lawson, K. D. & Flautt, T. J. (1968) J. Phys. Chem. 72, 2066–2074
4. Forbes, J., Husted, C. & Oldfield, E. (1988) J. Am. Chem. Soc. 110, 1059–1065
5. Forbes, J., Bowers, J., Moran, L., Shan, X., Oldfield, E. & Moscarello, M. A. (1988) J. Chem. Soc., Faraday Transactions 84, 3821–3849
6. Wennerström, H. (1973) Chem. Phys. Lett. 18, 41–44
7. Koenig, S. H., Brown III, R. D., Spiller, M. & Lundbom, N. (1990) Mag. Res. Med. 14, 482–495
8. Fralix, T. A., Ceckler, T. L., Wolff, S. D., Simon, S. A. & Balaban, R. S. (1991) Mag. Res. Med. 18, 214–223
9. Vist, M. R. & Davis, J. H. (1990) Biochemistry 29, 451–464
10. Oldfield, E., Chapman, D. & Derbyshire, W. (1971) FEBS Lett. 16, 102–104
11. Seelig, J. (1977) Quart. Rev. Biophys. 10, 353–418
12. Smith, R. L. & Oldfield, E. (1984) Science 225, 280–287
13. Davis, J. H. (1989) Adv. Magn. Res. 13, 195–223
14. Griffin, R. G. (1981) Methods Enzymol. 72, 108–174
15. Davis, J. H., Jeffrey, K. R., Bloom, M., Valic, M. I. & Higgs, T. P. (1976) Chem. Phys. Lett. 42, 390–394
16. Whittall, K. P., Sternin, E., Bloom, M. & MacKay, A. L. (1989) J. Magn. Res. 84, 64–71
17. Press, W. H., Flannery, B. P., Teukolsky, S. A. & Vetterling, W. T. (1986) in Numerical Recipes: the Art of Scientific Computing, pp. 523–528, Cambridge University Press, New York
18. Kang, S. Y., Gutowsky, H. S., Hsung, J. C., Jacobs, R., King, T. E., Rice, D. & Oldfield, E. (1979) Biochemistry 18, 3257–3267
19. Paddy, M. R., Dahlquist, F. W., Davis, J. H. & Bloom, M. (1981) Biochemistry 20, 3152–3162
20. Seelig, A. & Seelig, J. (1978) Hoppe-Seyler's Z. Physiol. Chemie 359, 1747–1756
21. Meier, P., Sachse, J. H., Brophy, P. J., Marsh, D. & Kothe, G. (1987) Proc. Natl. Acad. Sci. U.S.A. 84, 3704–3708

22. Rajan, S., Kang, S. Y., Gutowsky, H. S. & Oldfield, E. (1981) J. Biol. Chem. **256**, 1160–1166
23. Seelig, J., Gally, H.-U. & Wohlgemuth, R. (1977) Biochim. Biophys. Acta **467**, 109–119
24. Skarjune, R. & Oldfield, E. (1982) Biochemistry **21**, 3154–3160
25. Jarrell, H. C., Jovall, P. Å., Giziewicz, J. B., Turner, L. A. & Smith, I. C. P. (1987) Biochemistry **26**, 1805–1811
26. Oldfield, E., Chapman, D. & Derbyshire, W. (1972) Chem. Phys. Lipids **9**, 69–81
27. Stockton, G. W., Johnson, K. G., Butler, K. W., Tulloch, A. P., Boulanger, Y., Smith, I. C. P., Davis, J. H. & Bloom, M. (1977) Nature (London) **269**, 267–268
28. Oldfield, E. & Chapman, D. (1971) Biochem. Biophys. Res. Commun. **43**, 610–616
29. Keough, K. M., Oldfield, E., Chapman, D. & Beynon, P. (1973) Chem. Phys. Lipids **10**, 37–50
30. Oldfield, E., Bowers, J. L. & Forbes, J. (1987) Biochemistry **26**, 6919–6923
31. Oldfield, E., Chung, J., Le, H., Bowers, T., Patterson, J. & Turner, G. L. (1992) Macromolecules **25**, 3027–3030
32. Adebodun, F., Chung, J., Montez, B., Oldfield, E. & Shan, X. (1992) Biochemistry **31**, 4502–4509
33. Rothwell, W. P. & Waugh, J. S. (1981) J. Chem. Phys. **74**, 2721–2732
34. VanderHart, D. L., Earl, W. L. & Garroway, A. N. (1981) J. Magn. Res. **44**, 361–401
35. Cater, B. R., Chapman, D., Hawes, S. M. & Saville, J. (1974) Biochim. Biophys. Acta **363**, 54–69
36. Pines, A., Gibby, M. G. & Waugh, J. S. (1972) J. Chem. Phys. **56**, 1776–1777
37. Nicholson, L. K., Moll, F., Mixon, T. E., LoGrasso, P. V., Lay, J. C. & Cross, T. A. (1987) Biochemistry **26**, 6621–6626
38. Creuzet, F., McDermott, A., Gebhard, R., van der Hoef, K., Spijker-Assink, M. B., Herzfeld, J., Lugtenburg, J., Levitt, M. H. & Griffin, R. G. (1991) Science **251**, 783–786
39. Prosser, R. S., Davis, J. H., Dahlquist, F. W. & Lindorfer, M. A. (1991) Biochemistry **30**, 4687–4696
40. Cornell, B. A., Separovic, F., Baldassi, A. J. & Smith, R. (1988) Biophys. J. **53**, 67–76
41. Bowers, J. L. & Oldfield, E. (1988) Biochemistry **27**, 5156–5161

Rotational diffusion of membrane proteins: studies of band 3 in the human erythrocyte membrane using triplet probes

Richard J. Cherry

Department of Chemistry and Biological Chemistry, University of Essex, Wivenhoe Park, Colchester CO4 3SQ, U.K.

Introduction

The first quantitative determination of the rotational diffusion of a membrane protein was reported in 1972 by Cone who exploited the long-lived transient absorbance changes of rhodopsin to measure its rotation by an elegant microspectro-photometric method [1]. At that time K. Razi Naqvi, J. Gonzalez-Rodriguez and I, working in Dennis Chapman's group at the University of Sheffield, were interested in developing a general method for measuring rotational diffusion of proteins in membranes. Because lipid bilayers were thought to be much more viscous than water, we expected that rotation of proteins in membranes would be much slower than in aqueous solution and this was born out by the relaxation time of 20 μs which Cone measured for rhodopsin in rod outer segment disc membranes. It was clear that the well established technique of fluorescence depolarization would be inappropriate for the microsecond time range and our approach was to try to develop an analogous method based on the long lifetime of the triplet state of a suitable probe molecule. Working as we were in the Department where Sir George Porter developed flash photolysis, it was particularly appropriate that our first instrument, which was built by K. Razi Naqvi, used this method for triplet-state detection.

Early experiments were carried out with a very simple model system, namely BSA dissolved in a viscous glycerol/water mixture. This enabled us to test a range of dyes as potential triplet probes simply by binding them non-covalently to the protein's hydrophobic site. Eventually we found that eosin worked very satisfactorily and we were able with this probe to measure microsecond rotations of the protein [2].

Further progress was facilitated by synthesis of a reactive derivative, eosin isothiocyanate [3]. With this new probe it was possible to first test the method thoroughly on a range of proteins in viscous solution [4] and then to obtain an approximate measurement of protein rotation in the human erythrocyte membrane [5].

Subsequently, many studies of protein rotational mobility in membranes using triplet probes have been reported and a number of reviews have been published [6-9]. Rotational mobility of integral membrane proteins in cell membranes is frequently found to be complex. Populations of proteins with correlation times corresponding to those expected for free diffusion in a fluid lipid bilayer often coexist with more slowly rotating or immobile species. Restrictions on free diffusion are very probably a consequence of protein–protein interactions in the membrane. Thus diffusion measurements can lead towards a greater understanding of the detailed structure of membranes. This is particularly the case in the erythrocyte membrane where knowledge of the biochemistry of the membrane proteins has permitted a more detailed interpretation of mobility measurements than has been possible with most other cell membranes. Thus, the erythrocyte membrane has served as an excellent system for developing strategies to elucidate protein–protein interactions on the basis of diffusion measurements. Here, the methodology of protein rotational diffusion measurements is reviewed briefly and applications to the human erythrocyte membrane are summarized and discussed.

Measurement of protein rotation with triplet probes

Optical spectroscopic methods of measuring rotational motion were first introduced by Perrin [10] and later developed by Weber [11] for application to proteins in aqueous solution. These methods depend on creating an anisotropic molecular orientation by the process of photoselection. When a randomly oriented population of chromophores is excited by linearly polarized light, these chromophores whose transition dipole moment for the selected absorption lies in or near to the direction of polarization are selectively excited. Thus, the anisotropic distribution of molecules in the excited state is photoselected from the initial random distribution. If the excitation is from a brief pulse of light, then the initial anisotropy created by the flash decays at a rate of determined by the rotational diffusion coefficient D_R. By monitoring the decay of anisotropy, the value of D_R can in principle be determined.

Variations of the method arise through the detection of different signals in order to measure the anisotropy. Most commonly, the anisotropy is measured by detecting polarization of fluorescent emission from the molecules in the excited state. However, rotational diffusion can only be measured if a significant rotation occurs during the lifetime of the detected signal. This restricts the classic fluorescence method to the study of rotations in the nanosecond time range (or faster), which is appropriate for proteins in aqueous solution but not for proteins in membranes where rotations occur in the microsecond or even millisecond time range.

Triplet states can have millisecond lifetimes at room temperature and hence provide suitably long-lived signals for measuring rotation of membrane proteins. Because excitation to the triplet state changes the molecule's absorption spectrum, anisotropic decay can be monitored by observing dichroism of the absorbance changes. Following the development of a suitable triplet probe, the method was first applied to band 3 proteins in the human erythrocyte membrane [5].

The above method involving detection of absorption signals is generally known as 'transient dichroism'. Suitable instrumentation for performing transient dichroism measurements has been described in some detail [6, 12, 13]. The technique works well for proteins that are relatively abundant in the membrane. The same instrumentation can also be used to measure transient dichroism of some intrinsic chromophores such as the retinal chromophore of bacteriorhodopsin [2, 14]. However, sensitivity is inherently limited because absorption methods require detection of a change in the intensity of transmitted light. Emission methods, in which the signal is detected against a zero background, offer the possibility of greater sensitivity. Emission methods of detecting the triplet state include phosphorescence and delayed fluorescence. Methods based on the polarization of phosphorescence have been developed and used extensively [6, 7, 15–18], whereas the use of delayed fluorescence has been reported only occasionally [7, 19, 20]. Virtually all measurements with the above methods employ pulse excitation, although steady-state phosphorescence polarization has also been used [21, 22]. Different methods of triplet state detection are summarized in Fig. 1.

A particularly promising variation, known as fluorescence depletion, was introduced by Johnson and Garland [23]. When molecules are excited into the triplet state there is a corresponding fall in the concentration of molecules in the ground state. In the transient dichroism method this is detected by a decrease in the intensity of the lowest singlet–singlet absorption band. Johnson and Garland showed that a corresponding decrease in fluorescence could be measured with a spectacular increase in sensitivity. The gain in sensitivity is such that microscopic measurements on single cells are feasible. Further developments of the fluorescence depletion technique have been reported [24, 25], including the proposal that still higher sensitivities may be obtained by employing phase modulation methods [26, 27]. A variation of fluorescence depletion in which the probe is permanently photobleached by polarized light may be employed for the measurement of slow rotational motion [28, 29].

Fig. 1. Electronic energy level diagram illustrating different methods of detecting molecules in the triplet state T_1

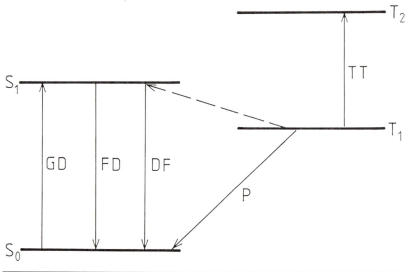

Phosphorescence (P) is emission accompanying a radiative transition back to the ground state S_0 whilst delayed fluorescence (DF) occurs when the molecule returns to S_0 via the excited singlet state S_1. Population of T_1 depletes the ground state resulting in loss of intensity of singlet–singlet absorption (GD; ground state depletion) and prompt fluorescence (FD; fluorescence depletion) under continuous illumination. Finally, molecules in the triplet state undergo triplet–triplet absorption (TT) when illuminated with light of appropriate wavelength. T_1 is populated by intersystem crossing from S_1 after exciting the S_0–S_1 transition. For simplicity, vibrational levels are omitted.

Data analysis

The time-dependent anisotropy $r(t)$ is defined by:

$$r(t) = \frac{A_\parallel(t) - A_\perp(t)}{A_\parallel(t) + 2A_\perp(t)} \tag{1}$$

where t is the time after excitation. For transient dichroism, $A_\parallel(t)$ and $A_\perp(t)$ are the absorbance changes for light polarized parallel and perpendicular, respectively, to the polarization of the exciting pulse. The same equation with absorbances replaced by intensities applies to phosphorescence and delayed fluorescence measurements. The anisotropy decay is normally determined only by rotational motion but complications can occur when the excited state decay is multiexponential.

Analysis of the anisotropy decay usually involves fitting $r(t)$ to a sum of exponential terms by a non-linear least squares procedure, i.e.:

$$r(t) = \sum_{i=1}^{n} r_i \exp(-t/\phi_i) + r_\infty \quad (2)$$

where ϕ_i are the rotational correlation times and r_∞ is the residual anisotropy. The initial anisotropy at $t=0$ is given by:

$$r_0 = \sum_{i=1}^{n} r_i + r_\infty \quad (3)$$

In most cases, the fit of eqn. (2) to the experimental data is not improved by including terms beyond $n=2$.

The relationship between correlation times and rotational diffusion coefficients and hence molecular parameters depends on an assumed model for molecular rotation [8, 30, 31]. For membrane proteins, the simplest model is that of uniaxial rotation in which the protein can rotate only about an axis normal to the plane of the membrane with rotational diffusion coefficient D_{11}. In this case:

$$r(t)/r_0 = A_1 \exp(-t/\phi_1) + A_2 \exp(-t/\phi_2) + A_3 \quad (4)$$

where

$$\phi_1 = 1/D_{11}, \; \phi_2 = 1/4D_{11}$$

The coefficients A_1, A_2 and A_3 are functions of the angle θ between the probe's transition dipole moment and the membrane normal. This model has been shown to be in good accordance with experimental data in the case of bacteriorhodopsin reconstituted into lipid vesicles [14].

Structure of the erythrocyte membrane

The erythrocyte cytoskeleton consists of a network of peripheral proteins which controls cell shape and provides the mechanical strength necessary to prevent rupture of the cell by the high shear forces experienced in blood capillaries. In recent years, the principal structural features of the cytoskeleton have been established mainly by studying the binding properties of isolated components [32, 33]. A schematic diagram, illustrating what are believed to be the principal features of the erythrocyte cytoskeleton, is shown in Fig. 2. Essentially, tetrameric units of spectrin are thought to be linked at junctions formed by actin and band 4.1. Two linking proteins, ankyrin (band 2.1) and band 4.1, provide the means by which the spectrin network attaches to the overlying cell membrane. Ankyrin, a 215 kDa polypeptide, possesses binding sites for spectrin and the cytoplasmic domain of band 3, which have been localized to 55 kDa acidic and 82 kDa basic tryptic fragments, respectively [34, 35].

Band 4.1 also has a membrane-binding site other than spectrin. Early studies have suggested this to be glycophorin A [36], although later

Fig. 2. **Schematic diagram illustrating principal interactions between cytoskeletal and integral proteins of the human erythrocyte membrane**

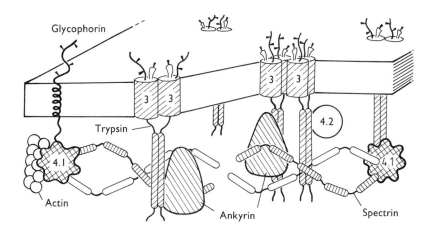

The trypsin-sensitive site on band 3 is also indicated. Proteins involved with the cytoskeleton not shown in the model include band 4.9, adducin and tropomysin.

work has shown this particular association to require the presence of a polyphosphoinisotide [37]. A lower-affinity lipid-independent site is available on band 3 [38] and attachment sites on the other glycophorins and phosphatidylserine have also been proposed.

Band 3 is a major integral membrane protein and comprises two domains which can express their function independently. Mild trypsin treatment produces a 43 kDa water-soluble cytoplasmic domain, leaving another domain of 55 kDa spanning the membrane [39]. The membrane-spanning domain facilitates the exchange of Cl^- for HCO_3^- across the erythrocyte membrane [40]. The cytoplasmic domain appears from microscopic and hydrodynamic studies to be highly extended [41] with an axial ratio greater than 10. This region binds rather promiscuously to glycolytic enzymes and haemoglobin [42]. The cytoplasmic domain also binds tightly both to 4.2, whose function is unknown, and to ankyrin [32, 33].

Rotational diffusion of band 3

Band 3 is selectively labelled by the triplet probe, eosin-5-maleimide, when the probe is reacted with intact erythrocytes [43]. It is thus possible to study the rotational mobility of this protein within the complex environment of the erythrocyte membrane. Initial transient dichroism experiments were only semi-quantitative and revealed that band 3 has significant rotational mobility in ghosts at 37 °C with a correlation time in the order of 0.5 ms [5]. Subsequently, more accurate data permitted analysis of the anisotropy decays into two exponential components plus a constant term [43]. A constant term is expected for uniaxial rotation (see eqn. 4) although immobile components can also contribute. The correlation times for the two mobile components were in the order of 150 μs and a few ms, respectively. These findings have been confirmed by phosphorescence anisotropy decay measurements [15, 44]. Recently, Matayoshi and Jovin [45] have resolved a third component with a correlation time of about 30 μs. They also succeeded in measuring rotational diffusion of band 3 in intact erythrocytes and found that it was essentially identical to that measured in ghosts. This is important because, previously, essentially all the detailed studies of band 3 mobility had been performed in ghosts. Much of the interest in band 3 rotational mobility has focused on elucidating the structural implications of the different correlation times.

Cytoskeletal interactions

There is rather convincing evidence that slowly rotating band 3 molecules, with correlation times of milliseconds or longer, are restricted by interaction with cytoskeletal proteins. Removal of the cytoplasmic domain of band 3 by mild proteolysis with trypsin results in a large decrease in the fraction of slowly rotating band 3 and a corresponding increase in the more rapidly rotating population [45, 46]. However, the molecular associations responsible for the restriction are complex and still not fully understood. It was shown early on that removal of spectrin and actin has no appreciable effect on band 3 rotation [5] and this has been confirmed [46, 47] in more detailed analyses. It is thus not surprising that Tsuji et al. [44] found that band 3 rotational mobility is independent of the spectrin dimer/tetramer equilibrium. After spectrin/actin depletion, subsequent removal of ankyrin and band 4.1 does enhance band 3 mobility in a manner qualitatively similar to removal of the cytoplasmic domain [46] (see Fig. 3). Yet *in situ* proteolysis of ankyrin does not have any observable effect on band 3 rotation [47].

These various observations can be partially reconciled with the model in Fig. 2 by consideration of the flexibility of the spectrin molecule [48]. Using transient dichroism measurements, it was shown recently that spectrin retains high flexibility even when forming part of the erythrocyte cytoskeleton [49]. Thus, the spectrin molecule is too floppy to impose a restriction on band 3 angular displacements over $\pm \pi/2$, even though it is linked to band 3 via ankyrin.

Although flexibility plausibly explains why spectrin does not influence the correlation times of band 3, the problem remains of determin-

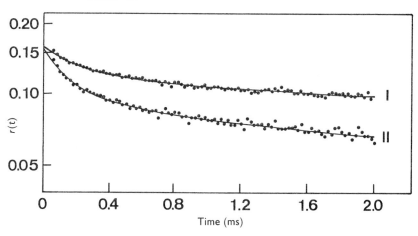

Fig. 3. Effect on the rotational mobility of band 3 of stripping ghosts of spectrin, actin, ankyrin and band 4.1

Shown are the anisotropy decay curves for eosin-labelled band 3 in control ghosts (I) and stripped ghosts (II). Measurements were made at 37 °C and pH 7.4. Reproduced from [46] with permission.

ing what exactly the interactions responsible for the slow component are. The fact that band 3 mobility is enhanced by removal of ankyrin and band 4.1 (subsequent to spectrin/actin depletion) shows one or both of these proteins must somehow be involved. Recently, we have studied this problem further by investigating the effects on band 3 rotational mobility of rebinding ankyrin and band 4.1 to ghosts stripped of these proteins, as well as spectrin and actin [50]. We find that rebinding either ankyrin or band 4.1 alone has no detectable effect on band 3 mobility. Rebinding both these proteins together does, however, reimpose a restriction on band 3 rotation (Fig. 4). The effect is not observed if the cytoplasmic domain of band 3 is removed with trypsin (Fig. 5). Some experiments were also performed with ghosts additionally depleted of band 4.2. The results of the ankyrin and band 4.1 rebinding experiments were similar, irrespective of the presence or absence of band 4.2.

The conclusion of these studies is that the slow rotational component of band 3 arises from interactions which require the presence of both ankyrin and band 4.1 on the membrane. Since rotational diffusion is rather sensitive to molecular size, it is very probable that ankyrin and band 4.1 promote the formation of small clusters of band 3 molecules, as suggested schematically in Fig. 6. It should be emphasized that the promotion of band 3 clustering is not predicted by the known binding properties of these proteins and the interactions responsible therefore remain to be elucidated.

Fig. 4. **Effect on the rotational mobility of band 3 of rebinding both ankyrin and band 4.1 to stripped ghosts**

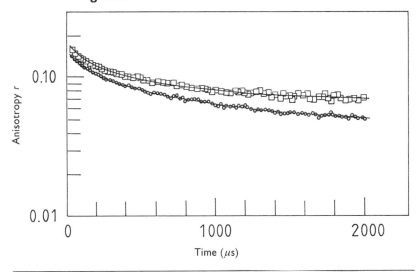

Shown here are the anisotropy decay curves for eosin-labelled band 3 in ghosts stripped of spectrin, actin, band 6, ankyrin and band 4.1 (○), and in stripped ghosts with ankyrin and band 4.1 rebound (□). Rebinding ankyrin or band 4.1 alone has no effect on band 3 rotational mobility. Measurements were made at 37°C and pH 7.4. Reproduced from [50] with permission.

Self-association of band 3

When the temperature of ghosts is reduced to below 37 °C, band 3 becomes increasingly immobile [43, 44]. As illustrated in Fig. 7, this is the case even if the cytoplasmic domain of band 3 is cleaved with trypsin to remove interactions with cytoskeletal proteins. Similar behaviour is observed when purified band 3 is reconstituted into lipid vesicles [51, 52]. It may be deduced that band 3 self-associates as a function of temperature. Self-association also appears to be promoted by the presence of cholesterol in the membrane [53].

A further question is what is the smallest oligomeric state of band 3 in the membrane? In principle, the diameter of the rotational species can be

Fig. 5. Effect on the rotational mobility of band 3 of rebinding both ankyrin and band 4.1 to trypsin-treated ghosts

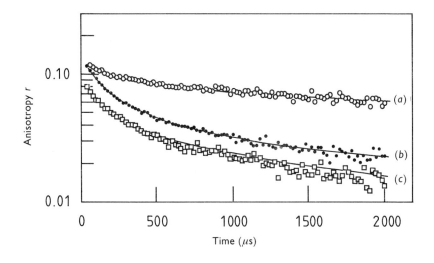

Shown here are the anisotropy decay curves for eosin-labelled band 3 in (a) control ghosts (b) trypsin-treated ghosts and (c) trypsin-treated ghosts incubated with ankyrin and band 4.1. For clarity, curve (c) has been vertically displaced. Measurements were made at 37°C and pH 7.4. Reproduced from [76] with permission.

obtained from the rotational diffusion coefficient D_R given by

$$D_R = kT/4\pi a^2 h\eta \tag{5}$$

where the protein is modelled as a cylinder of radius a spanning a membrane of thickness h and viscosity η [54]. Unfortunately, even a single D_R gives two correlation times and exact analysis of anisotropy decay curves becomes virtually impossible when multiple rotating species are present. Even in reconstituted systems it was not possible to eliminate the presence of aggregated band 3 [51]. However, Morrison et al. [55], using a 'global analysis' method, were able to deduce a value of $\phi_R = 1/D_R$ of 30 μs for the fastest component of band 3 rotation in lipid vesicles. This is quite similar to the fastest correlation time (28 μs) observed by Matayoshi and Jovin [45] for band 3 in ghosts. These measurements are consistent with band 3 existing as a dimer; however, because of uncertainty in the membrane viscosity and the shape of band 3, a tetramer cannot be ruled out.

Fig. 6. Schematic diagram of small clusters of band 3 induced by ankyrin and band 4.1 binding

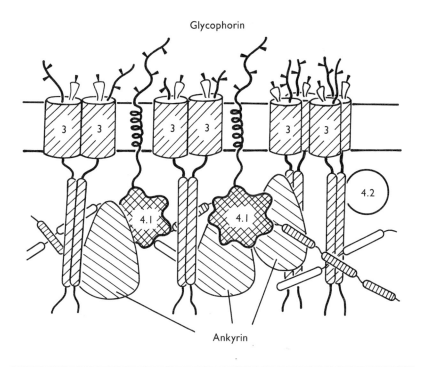

Details of the molecular interactions which promote clustering remain to be elucidated.

Other interactions of band 3

The possibility of band 3 and glycophorin A forming a complex in the erythrocyte membrane was examined by crosslinking glycophorin A with specific antibodies. This was found to immobilize band 3, suggesting that these two components are associated in the membrane [56].

The enzyme glyceraldehyde-3-phosphate dehydrogenase (GAPDH) is known to bind to the cytoplasmic domain of band 3. Removal of endogenous GAPDH from ghosts results in a relatively small decrease in the proportion of immobile band 3. Incubation of ghosts with GAPDH causes a progressive increase in the proportion of immobile band 3 such that, at saturating concentrations, nearly all of band 3 is immobile [57]. A similar though less complete effect was observed with aldolase, which also binds to the cytoplasmic domain of band 3. Although the mechanism of these effects is unclear, it is suggested that the enzymes promote crosslinking of band 3 into aggregates [57].

Melittin is a basic 26 amino acid polypeptide which causes haemolysis of erythrocytes. It also aggregates band 3, as shown by rotational diffu-

Fig. 7. **Temperature dependence of the anisotropy decay of eosin-labelled band 3 in which the cytoplasmic domain has been removed by mild proteolysis with trypsin**

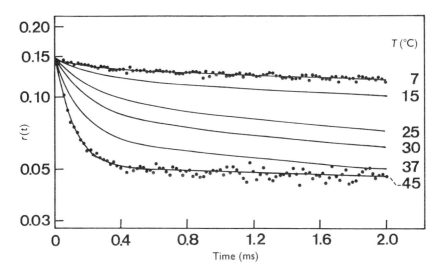

Measurements were made on ghosts at pH 7.4. The experimental points are omitted on intermediate curves for clarity. Reproduced from [46] with permission.

sion [58] and electron microscopy [59] studies. There is circumstantial evidence that protein aggregation is involved in melittin's haemolytic activity. Other basic proteins and polypeptides as well as divalent cations also aggregate band 3 [45, 47]. A comparison of different aggregating agents suggested that they fall into two groups. The first group consists of hydrophilic agents which are non-haemolytic and probably bind electrostatically to superficial anionic sites on band 3. The second group consists of haemolytic agents like melittin with a hydrophobic moiety which anchors them to lipid bilayers. They probably interact with anion sites close to the membrane surface and may also bind hydrophobically to transmembrane protein domains [47].

Tilley *et al.* [18] have examined band 3 rotational mobility in erythrocytes infected with the malarial parasite *Plasmodium falciparum*.

They find that band 3 is significantly immobilized in these infected cells. It was suggested that proteins of malarial origin may be crosslinking band 3 into aggregates, which may in turn be related to electron dense protruberances seen on the surface of infected cells. In other studies, Tilley et al. [60] observed decreased rotational mobility of band 3 in melanesian ovalocytes. These red cells have a mutant form of band 3 and are characterized by a high degree of membrane rigidity. However, the relationship between the mutation, the mobility of band 3 and membrane rigidity requires further elucidation.

Lateral diffusion

Fluorescence recovery after photobleaching (FRAP) is a well established technique for studying the lateral mobility of membrane components [7, 61-63]. In its simplest form, the method consists of photobleaching a small spot (typically 1 μm in diameter) on a membrane or cell surface with a brief intense burst of light from a focused laser beam. Subsequently fluorescence recovery, arising from diffusion of unbleached molecules into the bleached area, is monitored with low intensity illumination. The rate of recovery is determined by the diffusion coefficient, whilst the extent of recovery reveals the presence of immobile species.

The first photobleaching measurements with a fluorescent probe were in fact performed with erythrocyte membranes [64]. In these experiments, intact human erythrocytes were labelled with fluorescein isothiocyanate (most of the label was on band 3) and the fluorescence distribution monitored after photobleaching probes in one half of an erythrocyte ghost. No redistribution of fluorescence from the unbleached to the bleached half of the ghost was observed, implying a low lateral mobility of the labelled proteins.

The first measurements of rotational mobility of band 3, although only approximate, indicated that a significant fraction of band 3 is relatively freely mobile [5]. To rationalize the rotational and lateral diffusion data, it was proposed that band 3 is entrapped within the spectrin/actin network such that lateral diffusion over distances of the order of 1 μm is strongly impeded. Subsequent lateral diffusion studies in erythrocyte membranes have supported this interpretation [65-69]. In particular, Sheetz et al. [65] examined lateral mobility of integral proteins in mouse spherocytes which are deficient in cytoskeletal proteins. They found that the diffusion coefficient is some 50 times greater than in normal erythrocytes and approaches values measured for proteins freely diffusing in model lipid bilayer systems. Tsuji and Ohnishi [69] cleaved the cytoplasmic domain of band 3 with trypsin and found that lateral diffusion was increased about eightfold. A model which incorporates many of the experimental measurements of both lateral and rotational diffusion of band 3 is provided by Tsuji et al. [44]. According to this model, a fraction of band 3 is immobilized by direct attachment to the cytoskeleton. The remainder undergo free rotational diffusion (although probably slowed by self-association), but are laterally restricted by barriers formed by cytoskeletal proteins. As discussed above, this model probably requires some revision to account for the more recent rotational diffusion data.

Concluding remarks

The human erythrocyte membrane has proved to be an excellent system for studying protein mobility. It is sufficiently complex to pose a variety of interesting questions, yet the membrane components are well enough characterized to permit interpretable biophysical measurements to be performed. Techniques established with the erythrocyte membrane are being applied increasingly to other membranes. Moreover, homologues of many erythrocyte cytoskeletal proteins have now been identified in a variety of cell types [70]. Information on molecular interactions of these proteins and their influence on protein mobility derived from experiments with erythrocytes is therefore likely to be of general relevance. Very recently, direct observation of the tracks of membrane-bound particles labelled with colloidal gold or fluorescent probes have been reported [71–75]. Some of these experiments provide evidence for the existence of small domains in the plasma membranes of nucleated cells which may well be formed by cytoskeletal structures similar to those of the erythrocyte membrane [73].

> *I am indebted to the numerous colleagues who have contributed to studies described in this article. Current work is supported by the S.E.R.C.*

References

1. Cone, R. A. (1972) Nature (London) New Biol. 236, 39–43
2. Razi Naqvi, K., Gonzalez-Rodriguez, J., Cherry, R. J. & Chapman, D. (1973) Nature (London) New Biol. 245, 249–251
3. Cherry, R. J., Cogoli, A., Oppliger, M., Schneider, G. & Semenza, G. (1976) Biochemistry 15, 3653–3656
4. Cherry, R. J. & Schneider, G. (1976) Biochemistry 15, 3657–3661
5. Cherry, R. J., Bürkli, A., Busslinger, M., Schneider, G. & Parish, G. R. (1976) Nature (London) 263, 389–393
6. Thomas, D. D. (1986) in Techniques for the Analysis of Membrane Proteins (Ragan, C. I. and Cherry, R. J., eds.), pp. 377–431, Chapman and Hall, London, New York
7. Jovin, T. M. & Vaz, W. L. C. (1989) Methods Enzymol. 172, 471–573
8. Cherry, R. J. (1979) Biochim. Biophys. Acta 559, 289–327
9. Cherry, R. J. (1991) in The Structure of Biological Membranes (Yeagle, P. L., eds.), pp. 507–537, CRC Press Inc., Boca Raton
10. Perrin, F. (1936) J. Phys. Radium 7, 1–11
11. Weber, G. (1953) Adv. Protein Chem. 8, 415–419
12. Cherry, R. J. (1978) Methods Enzymol. 54, 47–61
13. Kinosita, K. Jr. & Ikegami, A. (1988) Subcell. Biochem. 13, 55–86
14. Cherry, R. J. & Godfrey, R. E. (1981) Biophys. J. 36, 257–276

15. Austin, R. H., Chan, S. S. & Jovin, T. M. (1979) Proc. Natl. Acad. Sci. U.S.A. **76**, 5650–5654
16. Moore, C., Boxer, D. & Garland, P. (1979) FEBS Lett. **108**, 161–166
17. Restall, C. J., Dale, R. E., Murray, E. K., Gilbert, C. W. & Chapman, D. (1984) Biochemistry **23**, 6765–6776
18. Tilley, L., Foley, M., Anders, R. F., Dluzewski, A. R., Gratzer, W. B., Jones, G. L. & Sawyer, W. H. (1990) Biochim. Biophys. Acta **1025**, 135–142
19. Razi Naqvi, K. & Wild, U. P. (1975) Chem. Phys. Lett. **36**, 222–224
20. Greinert, R., Stärk, H., Stier, A. & Weller, A. (1979) J. Biochem. Biophys. Methods **1**, 77–83
21. Strambini, G. B. & Galley, W. C. (1976) Nature (London) **260**, 554–556
22. Murray, E. K., Restall, C. J. & Chapman, D. (1983) Biochim. Biophys. Acta **732**, 347–351
23. Johnson, P. & Garland, P. B. (1981) FEBS Lett. **132**, 252–256
24. Yoshida, T. M. & Barisas, B. G. (1986) Biophys. J. **50**, 41–53
25. Corin, A. F., Blatt, E. & Jovin, T. M. (1987) Biochemistry **26**, 2207–2217
26. Garland, P. B. & Birmingham, J. J. (1986) Biochem. Soc. Trans. **14**, 838–839
27. Yoshida, T. M., Zarrin, F. & Barisas, B. G. (1988) Biophys. J. **54**, 277–288
28. Smith, L. M., Weis, R. M. & McConnell, H. M. (1981) Biophys. J. **36**, 73–91
29. Velez, M. & Axelrod, D. (1988) **53**, 575–591
30. Lipari, G. & Szabo, A. (1980) Biophys. J. **30**, 489–506
31. Kawato, S. & Kinosita, K. Jr. (1981) Biophys. J. **36**, 277–296
32. Bennett, V. (1985) Annu. Rev. Biochem. **54**, 273–304
33. Bennett, V. (1989) Biochim. Biophys. Acta **988**, 107–121
34. Weaver, D. C., Pasternak, G. R. & Marchesi, V. T. (1984) J. Biol. Chem. **259**, 6170–6175
35. Wallin, R., Culp, E., Coleman, D. & Goodman, S. R. (1983) Proc. Natl. Acad. Sci. U.S.A. **81**, 4095–4099
36. Anderson, R. A. & Lovrien, R. E. (1984) Nature (London) **307**, 655–658
37. Anderson, R. A. & Marchesi, V. T. (1985) Nature (London) **318**, 295–298
38. Pasternak, G. R., Anderson, R. A., Leto, T. L. & Marchesi, V. T. (1985) J. Biol. Chem. **260**, 3676–3683
39. Steck, T. L., Ramoz, B. & Strapezon, E. (1976) Biochemistry **15**, 1154–1161
40. Cabantchik, Z. I., Knauf, P. A. & Rothstein, A. (1978) Biochim. Biophys. Acta **515**, 239–302
41. Low, P. S., Westfall, M. A., Allen, D. P. & Appell, K. C. (1984) J. Biol. Chem. **259**, 13070–13076
42. Steck, T. L. (1978) J. Supramol. Struct. **8**, 311–324
43. Nigg, E. A. & Cherry, R. J. (1979) Biochemistry **18**, 3457–3538
44. Tsuji, A., Kawasaki, K., Ohnishi, S., Merkle, H. & Kusumi, A. (1988) Biochemistry **27**, 7447–7452
45. Matayoshi, E. D. & Jovin, T. M. (1991) Biochemistry **30**, 3527–3538
46. Nigg, E. A. & Cherry, R. J. (1980) Proc. Natl. Acad. Sci. U.S.A. **77**, 4702–4706
47. Clague, M. J., Harrison, J. P. & Cherry, R. J. (1989) Biochim. Biophys. Acta **981**, 43–50
48. Learmonth, R. P., Woodhouse, A. G. & Sawyer, W. H. (1989) Biochim. Biophys. Acta **987**, 124–128
49. Clague, M. J., Harrison, J. P., Morrison, I. E. G., Wyatt, K. & Cherry, R. J. (1990) Biochemistry **29**, 3898–3904
50. Wyatt, K. & Cherry, R. J. (1992) Biochim. Biophys. Acta **1103**, 327–330
51. Mühlebach, T. & Cherry, R. J. (1985) Biochemistry **24**, 975–983
52. Dempsey, C. E., Ryba, N. J. P. & Watts, A. (1986) Biochemistry **25**, 2180–2187.
53. Mühlebach, T. & Cherry, R. J. (1982) Biochemistry **21**, 4225–4228
54. Saffman, P. G. & Dulbrück, M. (1975) Proc. Natl. Acad. Sci. U.S.A. **72**, 3111–3113
55. Morrison, I. E. G., Mühlebach, T. & Cherry, R. J. (1986) Biochem. Soc. Trans. **14**, 885–886
56. Nigg, E. A., Bron, C., Girardet, M. & Cherry, R. J. (1980) Biochemistry **19**, 1887–1893
57. Matayoshi, E. D., Sawyer, W. H. & Jovin, T. M. (1991) Biochemistry **30**, 3539–3543
58. Dufton, J. J., Hider, R. C. & Cherry, R. J. (1984) Eur. Biophys. J. **11**, 17–24
59. Hui, S. W., Stewart, C. M. & Cherry, R. J. (1990) Biochim. Biophys. Acta **1023**, 335–340
60. Tilley, L., Nash, G. B., Jones, G. L. & Sawyer, W. H. (1991) J. Membr. Biol. **121**, 59–66
61. Axelrod, D., Koppel, D. E., Schlessinger, J., Elson, E. & Webb, W. W. (1976) Biophys. J. **16**, 1055–1069
62. Jacobson, K., Derzko, Z., Wu, E.-S., Hou, Y. & Poste, G. (1976) J. Supramol. Struct. **5**, 565–576
63. Peters, R. (1991) in New Techniques of Optical Microscopy and Microspectroscopy (Cherry, R. J., ed.), pp. 199–228, Macmillan Press, Basingstoke
64. Peters, R., Peters, J., Tews, K. H. & Bähr, W. (1976) Biochim. Biophys. Acta **367**, 282–294
65. Sheetz, M. P., Schindler, M. & Koppel, D. E. (1980) Nature (London) **285**, 510–512
66. Schindler, M., Koppel, D. E. & Sheetz, M. P. (1980) Proc. Natl. Acad. Sci. U.S.A. **77**, 1457–1461

67. Koppel, D. E., Sheetz, M. P. & Schindler, M. (1981) Proc. Natl. Acad. Sci. U.S.A. 78, 3576–3580
68. Golan, D. E. & Veatch, W. (1980) Proc. Natl. Acad. Sci. U.S.A. 77, 2537–2541
69. Tsuji, A. & Ohnishi, S. (1986) Biochemistry 25, 6133–6139
70. Bennett, V. (1990) Physio. Rev. 70, 1029–1065
71. Sheetz, M. P., Turney, S., Qian, H. & Elson, E. L. (1989) Nature (London) 340, 284–288
72. de Brabander, M., Nuydens, R., Ishihara, A., Holifield, B., Jacobson, K. & Geerts, H. (1991) J. Cell Biol. 112, 111–124
73. Edidin, M, Kuo, S. C. & Sheetz, M. P. (1991) Science 254, 1379–1382
74. Gross, D. J. & Webb, W. W. (1988) in Spectroscopic Membrane Probes, Vol. II (Loew, L. M., ed.), pp. 19–45, CRC Press, Boca Raton
75. Anderson, C. M., Georgiou, G. N., Morrison, I. E. G., Stevenson, G. V. & Cherry, R. J. (1992) J. Cell Sci. 101, 415–425
76. Wyatt, K. (1990) Ph.D. Thesis, University of Essex

Modulation of protein structure by the lipid environment

Witold K. Surewicz, Arturo Muga and Henry H. Mantsch

Institute for Biological Sciences (WKS) and Steacie Institute for Molecular Sciences (HHM) of the National Research Council of Canada, K1K-0R6, Canada and Department of Biochemistry, University of the Basque Country, Bilbao, Spain

Introduction

The two major components of biological membranes are lipids and proteins. While the structural and dynamic properties of lipids are relatively well characterized, the understanding of the structure of membrane proteins is at a much less advanced level. Current models of protein folding in a membrane environment are based largely on theoretical considerations (e.g., hydrophobicity plots) derived from amino acid sequences. Unfortunately, progress in determining protein primary structure has not yet been matched by direct experimental determinations of membrane protein structure at the secondary, tertiary and topological levels.

The poor understanding of the folding and conformational structure of membrane proteins appears to be largely a consequence of experimental difficulties in studying proteins in a lipid environment. The currently available high resolution techniques have been developed to probe the structure of globular, water-soluble proteins and are of limited value when applied to membrane proteins. Thus, membrane proteins are notoriously difficult to crystallize, and the crystallographic analysis of such proteins is still in its infancy. The two-dimensional high resolution n.m.r. techniques, which offer an increasingly viable alternative for studying water-soluble proteins, are of limited use because of motional restrictions of proteins in a membrane environment. Difficulties are also encountered in probing membrane-protein structure by optical spectroscopic methods such as circular dichroism and (to a lesser extent) fluorescence spectroscopy. The light scattering effect on large membrane fragments can distort optical spectra, leading to errors in their interpretation. The technique which emerged recently as a valuable probe of membrane-protein structure is Fourier-transform i.r. spectroscopy [1–4]. However, the information provided by i.r. spectroscopy is usually at a low level of structural detail and the current methods of interpretation of protein vibrational spectra require further refinement.

The experimental difficulties encountered in structural and conformational studies apply not only to native membrane proteins, but also to a much larger class of proteins (and peptides) which, although soluble in water, during the course of their action interact with plasma or intracellular membranes [5]. The association–dissociation equilibria of such transient membrane-protein complexes are often controlled by the metabolic state of the cell. The changes in protein or peptide conformation caused by the association and interaction with the membrane surface (or its interior) are believed to be instrumental in a large number of physiologically important processes. Examples of such processes include post-translational membrane translocation of newly synthesized proteins, interaction between peptide hormones and their target receptors, intoxication by bacterial toxins, assembly of the membrane attack complement complex, and fusion between virus envelopes and host cell membranes.

In this chapter, we present selected examples of membrane-protein/peptide interactions. These examples include interactions with peptide hormones, bacterial toxins and cytochrome c. We describe the specific conformational changes that a membrane environment can induce in proteins/peptides and discuss the potential biological implications of these conformational transitions.

Lipid-induced peptide folding: peptide hormones

The importance of peptide conformation in the interaction between peptide hormones and their receptors has been recognized for a long time. However, the conformational analysis of peptide hormones and the elucidation of structure–function relationships for these peptides present a number of complex and challenging problems [6]. Many biologically active peptides consist of relatively short sequences of amino acids. Such small molecules are generally highly flexible; in aqueous solutions they are likely to exist as an ensemble of conformers. Moreover, the conformation of the peptides is environment-sensitive; it may be affected by factors such as type of solvent, degree of self-association and binding to other molecules.

The ability of many peptide hormones to change conformation in response to environmental factors poses the question of whether the models of peptide conformation developed for the free molecules in aqueous solution or for peptide crystals, describe adequately the conformation prevailing at or near the receptor site. Although the notion that a 'bioactive' conformation of an approaching hormone is necessary for receptor recognition remains a matter of discussion, current concepts emphasize the importance of an interfacial environment and postulate a specific role for lipid affinity and lipid-induced peptide refolding in peptide hormone–receptor interactions [7, 8]. The postulated functional role of the lipid phase of the target cell membrane is (i) to facilitate accumulation of the peptide in a membrane compartment near the receptor site, (ii) to guide the peptide to the receptor by reducing a three-dimensional random search to a two-dimensional diffusion on the membrane, (iii) to orient the peptide in a way facilitating productive interaction with the receptor, and (iv) to induce the specific folded conformation of the peptide that meets the requirements of the receptor.

A structural feature common to many peptide hormones is the presence of sequences in which hydrophobic amino acid residues are spaced regularly along the peptide chain at every third or fourth residue. Such sequences, when placed in an interfacial environment, have a tendency to form amphipathic helices. In this conformation, hydrophobic and hydrophilic residues are spacially segregated on opposite faces of the helix. The potential for amphipathic helices in different sequences can be assessed by the use of helical wheel projections or, more quantitatively, by calculations of the hydrophobic moment [9]. A notable characteristic of amphipathic helix-forming peptides is their ability to bind to phospholipids and other biological interfaces [10]. Such binding is accompanied by a substantial increase in the helical content.

Examples of peptide hormones that, upon lipid binding, form amphipathic helices include glucagon [11], β-endorphin [12] and calcitonin [13–15]. The structural properties of the latter peptide have been explored in considerable detail. The resulting model postulates that the biological active structure of calcitonin contains three regions [14] — a seven-residue N-terminal 'active site', an amphipathic helix in the region of residues 8–22, and a hydrophilic C-terminal region containing residues 23–32. Although some analogues of calcitonin that cannot form an amphipathic helix were found to be inactive, no straightforward correlation between the α-helix hydrophobic moment and biological activity was found [15].

Another example of a peptide which, upon membrane binding, undergoes a transition from an unordered structure to an α-helix, is the bioactive fragment of adrenocorticotropic hormone, $ACTH_{1-24}$. The i.r. spectrum of the peptide in a membrane environment was found to be characteristic of a partly helical structure, with a preferential orientation of the helix axis perpendicular to the bilayer plane. In contrast, neither the N-terminal segment $ACTH_{1-10}$ nor the C-terminal fragment $ACTH_{11-24}$ alone were capable of being adsorbed spontaneously to the lipid membranes [16, 17]. The i.r. spectroscopic data provided the basis for a model of regioselective, conformation-selective and orientation-selective interaction of $ACTH_{1-24}$ with phospholipid membranes [16, 17]. This interaction is driven by the amphipathic primary structure and results in the membrane insertion of the orientated α-helical 'message' segment ($ATCH_{1-10}$), with the 'address' segment ($ATCH_{11-24}$) remaining on the membrane surface.

Another type of membrane interaction was found for the five-amino acid residue opioid peptides [Met-5]-enkephalin and [Leu-5]-enkephalin. Essentially the whole structure of these hormones constitutes the specific recognition site. The n.m.r. experiments of Deber et al. [7, 18] revealed that enkephalins associate with micelles of lysolecithin with a K_a value of around 3×10 M^{-1}. Whereas in water the peptides exist as an ensemble of largely unfolded conformers, the less polar micellar environment induces conformational transition to intramolecularly H-bonded β-turn structure. A similar conformational transition was observed using i.r. spectroscopy for enkephalins bound to bilayer membranes of phosphatidylcholine [19].

Although amphipathic helices are the most common structures of membrane-bound peptide hormones, some peptides show a distinctly higher porosity to form β-strands. An example of the latter class of hormones is the atrial natriuretic peptide [20]. The i.r. spectrum in aqueous solution of the 24-amino acid atriopeptin III consists of only one component band in the amide I region, at around 1643 cm^{-1}, which is characteristic of unordered conformations (Fig. 1). Unlike some other peptide hormones, atriopeptin III does not interact with neutral phosphatidylcholine membranes. However, the peptide binds strongly to membranes of acidic phospholipids. This electrostatic binding results in a substantial conformational change, as indicated by the drastic alteration of its i.r. spectrum (Fig. 1). The appearance of amide I component bands at 1618 and 1637 cm^{-1} allows us to infer that the structure of the membrane-bound peptide consists largely of β-strands.

A lipid-induced transition to a β-structure was also found for the eight-amino-acid natural pressor agent angiotensin II [21]. Furthermore, i.r. spectroscopic experiments with the N-truncated analogue of the hormone, heptapeptide des-Asp[1]-angiotensin II, revealed that, although the two peptides have a very similar conformation in aqueous solution, their structures in a membrane environment are markedly different. In contrast to angiotensin II, the lipid-bound truncated peptide shows very little tendency to form β-structure. This is somewhat unexpected since the two analogues have a comparable receptor affinity and similar biological activities. These differences in conformation of membrane-bound angiotensin II analogue point to the complex relationship that exists between the lipid affinity and the lipid-induced folding of peptide hormones and their biological acitivity.

Specific binding to the membrane receptor: cholera toxin

The initial step in the cytotoxic action of many bacterial toxins is binding to specific receptors on the surface of susceptible cells. The receptor-binding steps is followed by towin internalization (as in the case of diphtheria toxin, *Pseudomonas aeruginosa* exotoxin, cholera toxin or Shigella toxin), or by its membrane insertion and assembly into a transmembrane channel (as for example with colicin or α-toxin of *Staphylococcus aureus*) [22]. Progress in

Fig. 1. Infrared spectra of atriopeptin III in solution (bottom) and upon complexation with dimyristoylphosphatidylglycerol at a lipid-to-peptide molar ratio of 10:1 (top)

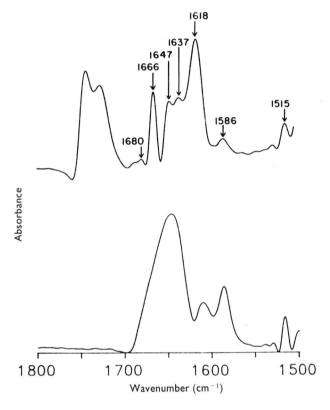

The bands between approximately 1720 and 1750 cm^{-1} represent C=O stretching vibrations of the lipid molecules. The original spectra were subjected to band-narrowing by Fourier self-deconvolution using a resolution enhancement factor of 2. Adapted from Surewicz et al. [20]

elucidating molecular aspects of the toxin–receptor interactions, as well as in understanding the complex mechanisms of the post-binding events, seems to be critically dependent on detailed conformational characterization of the toxin molecules both in solution and in their membrane-associated forms. Yet, direct spectroscopic studies of the toxin conformation at the receptor site are usually impractical because of the lack of availability of purified receptors and/or interference from the spectroscopic signal of the proteinaceous receptor molecule. In this context, unique opportunities are offered by cholera toxin. Unlike with many other toxins, the functional receptor for cholera toxin is not another protein but a simple glycolipid molecule, ganglioside G_{M1} [23]. This feature renders the receptor binding and the post-binding internalization of cholera toxin amenable to detailed spectroscopic and biophysical studies.

Cholera toxin is composed of two structurally and functionally distinct subunits, A and B [23]. All the toxic activity is associated with the A subunit. The role of the B subunit is to initiate the toxin-cell interaction by binding to ganglioside G_{M1} on the cell membrane surface. Structurally, this subunit forms a ring of five identical non-covalently bound polypeptide chains [22, 24].

There has been considerable interest in exploring structural and conformational aspects of cholera toxin–receptor interaction and subsequent internalization of the A subunit. Fluorescence studies have indicated that association with the receptor results in substantial changes in the microenvironment of the single tryptophan residue of the B subunit (Trp-88) [25, 26]. This, as well as other considerations, prompted speculations that association with the receptor brings about conformational changes in the B subunit which, in turn, provide the necessary driving force for the membrane translocation of the toxic A subunit. One model even assumes that, upon receptor binding, the B subunit enters the lipid bilayer and forms a hydrophilic channel through which the A component can diffuse [27]. However, such a transition from water-soluble to a membrane-soluble form would require a major refolding of the B subunit.

Recently, using i.r. spectroscopy, we have probed the secondary structure of cholera toxin in different environments [28]. Somewhat surprisingly, we found that the amide I bands of the B subunit in solution and in the receptor-bound form are very similar (Fig. 2). In both cases, the major component of the secondary structure was found to be β-sheet (bands at 1632 and 1671 cm^{-1} in Fig. 2b). It was estimated that no greater than 3–4% change in β-sheet or α-helix could occur upon B subunit binding to ganglioside G_{M1} receptor. These data clearly argue against any extensive change in the backbone conformation of the B subunit caused by binding to the surface receptor. The changes in the fluorescence spectrum thus are likely to reflect small and highly localized structural rearrangements in the vicinity of Trp-88 and/or a direct involvement of this residue in the receptor binding. Indeed, in the crystal structure of a cholera-toxin-related protein, Trp-88 was found at the bottom of a small cavity that probably forms part of the ganglioside-binding site [24].

Recent calorimetric experiments [29, 30] indicate that the major effect of ganglioside G_{M1} is on the cooperative interactions within the B subunit pentamer. The increased interactions between individual chains observed in the presence of the receptor may facilitate the release of the A subunit into the membrane.

An overall picture emerging from the biophysical studies is that receptor binding does not result in any major refolding of the B subunit, although it may increase communication between individual chains within the B pentamer. The main role of the B subunit seems to be to provide a

Fig. 2   Infrared spectra of the cholera toxin B subunit

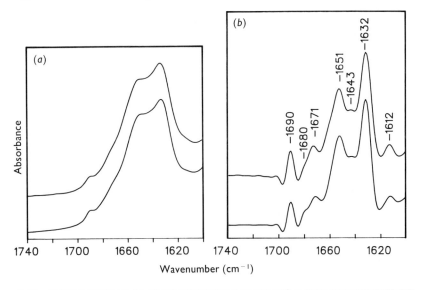

Part (a) shows the spectra of the subunit free in aqueous solution (bottom spectrum) and in the presence of ganglioside G_{M1} at the glycolipid-to-B-subunit molar ratio of 2:1 (top spectrum). Part (b) shows the same spectra after band narrowing by Fourier self-deconvolution. Reproduced from Surewicz et al. [28] with permission.

water-soluble carrier for the A subunit and to place this subunit in close proximity to the membrane surface in a manner facilitating its release into the membrane. The actual mechanism of membrane translocation of the toxic A subunit is likely to take advantage of the intrinsic properties of this subunit, such as its hydrophobicity and conformational flexibility.

Membrane-induced protein destabilization: cytochrome c

Cytochrome c, a 104-amino-acid protein, is an essential component of the respiratory chain in mitochondria. Its physiological role is to transfer electrons between the mitochondrial enzyme complexes cytochrome reductase and cytochrome oxidase; however, the detailed mechanism of this electron transfer reaction is far from being understood. The dynamic function of cytochrome c requires that the protein remains in transient contact and interacts with various components of the inner mitochondrial membrane. In model membrane systems, cytochrome c has been shown to interact strongly with acidic phospholipids and in particular with cardiolipin [31-33]. Various characteristics of this interaction allow cytochrome c to be classified as an extrinsic membrane protein.

The structural properties of cytochrome c in aqueous solution and as single crystals have been characterized in great detail by numerous spectroscopic methods and X-ray crystallography. Studies on cytochrome c–membrane interactions have focused largely on the effect of the protein on the conformation and molecular organization of membrane lipids. A salient assumption of these studies was that, upon membrane binding, the protein retains its well described solution (or crystalline) structure. Only recently, systematic studies have been undertaken to explore structural and thermodynamic properties of cytochrome c in physiologically relevant environments. From these studies an exciting picture emerges of complex structural transitions in the protein in response to electrostatic binding to negatively charged interfaces.

Initial efforts to characterize the conformational state of membrane-bound cytochrome c focused on the prosthetic haem group, revealing that binding to acidic phospholipids induces changes in the porphyrin ring conformation and haem coordination [34-38]. In a series of elegant resonance Raman experiments, Hildebrandt and colleagues [36-38] identified two stable conformational states of cytochrome c, at charged interfaces, that differ with respect to the coordination pattern of the haem iron and the structure of the haem pocket. In state I, the structure of the haem group and its environment remains essentially unchanged relative to the native cytochrome c in solution. State II, on the other hand, is characterized by a weakened axial Met–Fe bond, which leads to a thermal equilibrium between five-coordinated high-spin and six-coordinated low-spin configurations. In state II the haem crevice assumes an open conformation compared with the closed structure in state I. These structural changes give rise to a large shift of the redox potential. The two conformational states exist in an equilibrium that is controlled by the distribution of charges within the electrical double layer of the interface; they can be converted into each other by varying the surface potential.

A question arises as to whether the perturbations of the prosthetic group that are induced by interfacial binding of cytochrome c represent a local effect only, or whether they reflect more extensive changes in the conformational structure and/or thermodynamic state of the protein backbone. To address this question, we have probed the structural properties and stability of membrane-bound cytochrome c by means of i.r. spectroscopy, circular dichroism and differential scanning calorimetry [39]. The combined i.r. and circular dichroism data indicated that association of cytochrome c with phosphatidylglycerol-containing membranes results in only slight, if any, perturbation of the protein secondary structure. However, upon membrane binding a considerable increase in accessibility of protein backbone groups to hydrogen–deuterium exchange was observed. Such an increased

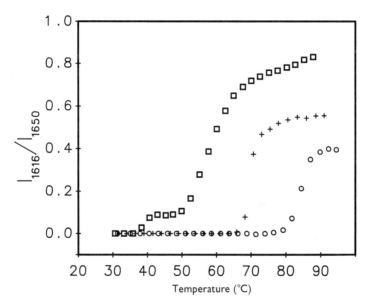

Fig. 3. Temperature dependence of the ratio of the amide I band intensity at 1616 cm^{-1} to that at 1650 cm^{-1} for ferricytochrome c in solution (○), ferricytochrome c associated with DMPG/DMPC (1:1) membranes (+), and ferricytochrome c associated with pure DMPG membranes (□)

The band at 1650 cm^{-1} is characteristic of the native protein structure, while the appearance of a band at 1616 cm^{-1} is associated with the thermal denaturation of the protein. Reproduced from Muga et al. [39] with permission.

solvent accessibility suggests lipid-induced loosening and/or destabilization of the protein tertiary structure. Consistent with this notion are results of thermal denaturation experiments. Thermotropic unfolding of cytochrome c in solution and in a lipid-bound form was studied by following changes in the i.r. amide I band (Fig. 3) and, more directly, by thermodynamic measurements with differential scanning calorimetry (Fig. 4). The major denaturation step of cytochrome c in aqueous buffer (pH 7.2) occurs at approximately 82°C. Upon binding to liposomes containing dimyristoylphosphatidylglycerol (DMPG) as a single lipid component, the denaturation temperature of cytochrome c decreases by approximately 30 degrees. This drastic drop in T_m is accompanied by a decrease of the calorimetric enthalpy of denaturation (from 94 kcal/M for the protein in solution to 24 kcal/

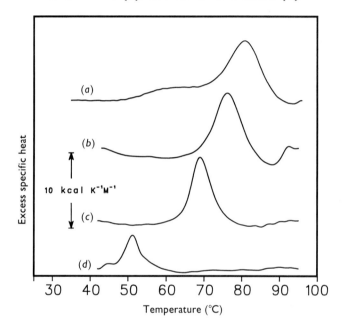

Fig. 4. Traces of differential scanning calorimetry scans for ferricytochrome c in buffer (a), for ferricytochrome c bound to membranes composed of a 1:3 mixture of DMPG and DMPC (b) or a 1:1 mixture of DMPG and DMPC (c), and for DMPG alone (d)

Reproduced from Muga et al. [39] with permission.

M for cytochrome c bound to DMPG). A destabilizing effect is also seen upon binding of cytochrome c membranes containing mixtures of acidic (DMPG) and zwitterionic (dimyristoylphosphatidylcholine; DMPC) phospholipids. Most notably, the extent of this structural destablization depends strongly on the surface density of negatively charged lipids: it decreases as the proportion of acidic DMPG in the membranes decreases.

The picture emerging from our studies is generally consistent with results of recent ^{31}P and deuterium n.m.r. experiments [40, 41] which provide evidence for reversible unfolding of cytochrome c upon interaction with bilayers of cardiolipin. Of particular significance is the observation that structural perturbations in the lipid-bound cytochrome c are modulated by the membrane surface potential. Electric field-induced structural perturbations could provide an effective mechanism for controlling the functional state of cytochrome c at different docking sites on the mitochondrial membrane. In this context, a specific model has been proposed recently [37] that postulates coupling of electric field-induced conformational changes within the haem group of cytochrome c with the electron transfer process in mitochondria.

References

1. Braiman, M. S. & Rothschild, K. J. (1988) Annu. Rev. Biophys. Biophys. Chem. 17, 541–570
2. Surewicz, W. K. & Mantsch, H. H. (1988) Biochim. Biophys. Acta 952, 115–130
3. Surewicz, W. K. & Mantsch, H. H. (1990) in Protein Engineering: Approaches to the Manipulation of Protein Folding (Narang, S., ed.), pp 131–157, Butterworth, Stoneham, M.A.
4. Chapman, D., Jackson, M. & Haris, P. I. (1989) Biochem. Soc. Trans. 17, 617–619
5. Martinek, K., Klyachko, N. L., Kabanov, A. V., Khmelnitsky, Y. L. & Levashov, A. V. (1989) Biochim. Biophys. Acta 981, 161–172
6. Blundell, T. & Wood, S. (1982) Annu. Rev. Biochem. 51, 123–154
7. Deber, C. M. & Behnam, B. A. (1985) Biopolymers 24, 105–116
8. Sargent, D. F. & Schwyzer, R. (1986) Proc. Natl. Acad. Sci. U.S.A. 83, 5774–5778
9. Eisenberg, D., Weiss, R. M. & Terwilliger, T. C. (1984) Proc. Natl. Acad. Sci. U.S.A. 81, 140–144
10. Kaiser, E. T. & Kezdy, F. J. (1987) Annu. Rev. Biophys. Biophys. Chem. 16, 561–581
11. Epand, R. M., Jones, A. J. S. & Schreier, S. (1977) Biochim. Biophys. Acta 491, 296–304
12. Wu, C.-S. C., Lee, N. M., Loh, K. H., Yang, J. T. & Li, C. H. (1979) Proc. Natl. Acad. Sci. U.S.A. 76, 3656–3659
13. Epand, R. M., Epand, R. F., Orlowski, R. C., Schlueter, R. J., Boni, L. T. & Hui, S. W. (1983) Biochemistry 22, 5074–5084
14. Moe, G., Miller, R. J. & Kaiser, E. T. (1983) J. Am. Chem. Soc. 105, 4100–4102
15. Epand, R. M., Seyler, J. K., Orlowski, R. C. (1986) Eur. J. Biochem. 159, 125–127
16. Gremlich, H. U., Fringeli, U. P. & Schwyzer, R. (1983) Biochemistry 22, 4257–4263
17. Gremlich, H. U., Fringeli, U. P. & Schwyzer, R. (1984) Biochemistry 23, 1808–1810
18. Behnam, B. A. & Deber, C. M. (1984) J. Biol. Chem. 259, 14935–14940
19. Surewicz, W. K. & Mantsch, H. H. (1988) Biochim. Biophys. Res. Commun. 150, 245–251
20. Surewicz, W. K., Mantsch, H. H., Stahl, G. L. & Epand, R. M. (1987) Proc. Natl. Acad. Sci. U.S.A. 84, 7028–7030
21. Surewicz, W. K. & Mantsch, H. H. (1988) J. Am. Chem. Soc. 110, 4412–4414
22. Stephen, J. & Pietrowski, R. A. (1980) Bacterial Toxins, American Society for Microbiology
23. Fishman, P. H. (1982) J. Membr. Biol. 69, 85–97
24. Sixma, T., Pronk, S. E., Kalk, K. H., Wartna, E. S., van Zanten, B. A. M., Witholt, B. & Hol, W. G. J. (1991) Nature (London) 351, 371–377
25. Moss, J., Osborne, J. C., Fishman, P. H., Brewer, H. B., Vaughan, M. & Brady, R. O. (1977) Proc. Natl. Acad. Sci. U.S.A. 74, 74–78
26. De Wolf, M. J. S., Fridkin, M. & Kohn, L. D. (1981) J. Biol. Chem. 256, 5489–5496
27. Gill, D. M. (1976) Biochemistry 15, 1242–1248
28. Surewicz, W. K., Leddy, J. J. & Mantsch, H. H. (1990) Biochemistry 29, 8106–8111
29. Goins, B. & Freire, E. (1988) Biochemistry 27, 2046–2052
30. Schon, A. & Freire, E. (1988) Biochemistry 28, 5019–5024
31. Rietveld, A., Sijens, P., Verkleij, A. J. & de Kruijff, B. (1983) EMBO J. 2, 907–913
32. Devaux, P. F., Hoatson, G. L., Favre, E., Fellmann, P., Farren, B., MacKay, A. L. & Bloom, M. (1986) Biochemistry 25, 3804–3812
33. Demel, R. A., Jordi, W., Lambrechts, H., van Damme, H., Hovius, R. & de Kruijff, B. (1989) J. Biol. Chem. 264, 3988–3997
34. Vincent, J. S. & Levin, I. W. (1986) J. Am. Chem. Soc. 108, 3551–3554
35. Vincent, J. S., Kon, H. & Levin, I. W. (1987) Biochemistry 26, 2312–2314
36. Hildebrandt, P. & Stockburger, M. (1989) Biochemistry 28, 6710–6721

37. Hildebrandt, P. & Stockburger, M. (1989) Biochemistry 28, 6722–6728
38. Hildebrandt, P., Heimburg, T. & Marsh, D. (1990) Eur. Biophys. J. 18, 193–201
39. Muga, A., Mantsch, H. H. & Surewicz, W. K. (1991) Biochemistry 30, 7219–7224
40. Spooner, P. J. R. & Watts, A. (1991) Biochemistry 30, 3871–3879
41. Spooner, P. J. R. & Watts, A. (1991) Biochemistry 30, 3880–3885

Infrared spectroscopic studies of membrane proteins

David C. Lee

SmithKline Beecham Pharmaceuticals, The Frythe,
Welwyn, Herts AL6 9AR, U.K.

Introduction

I.r. spectroscopy has been used extensively in the study of both lipid and protein components of biological membranes for many years. While the structural and dynamic properties of membrane lipids have been well characterized by many other biophysical techniques, i.r. spectroscopy is particularly well suited for investigations of the structure of membrane proteins, as it is insensitive to many of the sampling difficulties experienced by other methods. Many of the significant developments in this area have been made in the laboratory of Dennis Chapman over the past decade. These advances and those of other workers in this field are highlighted in this chapter.

Our knowledge of the structure and dynamics of membrane proteins is significantly less detailed than that of membrane lipids. This is a direct consequence of the difficulties in applying most of the established techniques for studying protein structure to systems which require the maintenance of an amphipathic environment for structural integrity. These problems [1] have severely restricted the application of single-crystal X-ray diffraction to relatively few membrane proteins (e.g. [2–4]). Although two-dimensional n.m.r. techniques are providing a great deal of structural detail about relatively low-molecular-weight proteins in solution, they have limited applicability to motionally restricted membrane proteins. Circular dichroism spectroscopy provides secondary structure data on soluble proteins, which can be compared to that available by i.r. However, light-scattering by membrane preparations can distort circular dichroism spectra leading to difficulties in their interpretation. The application of Raman spectroscopy to protein structure has been limited by sample fluorescence and relatively low signal-to-noise ratios.

At present we know relatively little of the folding of proteins within the lipid bilayer matrix. Studies based on primary sequence data frequently use an analysis of the hydrophobicity [5] along the polypeptide chain to predict transmembrane helices. This approach has been used to predict the membrane disposition of a large number of integral membrane proteins [6]. Alternative predictive methods rely on the secondary structure propensities of residues in a known sequence [7]. However, as propensities are based on structures of soluble proteins their application to transmembrane segments is questionable. An investigation of the accuracy of predictive methods compared with available spectroscopic and diffraction data has concluded that they are inappropriate for deducing the folding of membrane proteins [8].

I.r. spectroscopy is applied readily to native or reconstituted membrane samples and protein spectra are not distorted by the properties of the lipid environment. Therefore, it is the structural technique most easily applied to membrane proteins and, with recent advances in data manipulation and interpretation it is providing an increasing amount of detail. The principal areas of interest are the analyses of secondary structure composition, orientation of proteins within the lipid bilayer and mechanistic details of protein function.

Protein secondary structure from i.r. spectra

The sensitivity of the i.r. spectrum to protein conformation is well established [9-13]. Much of this early work concentrated on the assignment of the so-called 'amide' bands that are characteristic of proteins and peptides. This nomenclature arises from the similarity between the i.r. absorption bands exhibited by secondary amides and those of the CONH grouping of proteins. Amide bands I, II, III, IV, V, VI and VII may not be described by simple atomic displacements but result from delocalized vibrations of the CONH group [13]. In addition to the amide bands, the spectra of polypeptides and proteins will, of course, contain bands originating from vibrations of the amino acid side-chains.

Amide bands I, II and III are used most commonly in the characterization of protein secondary structure and accessibility to solvent. Ideally our interpretation of the i.r. spectrum in terms of secondary structure should be based on the results of normal mode calculations for defined structures. Remarkably good agreement has been found between the calculated frequencies of these bands and those calculated for simple peptides and homopolypeptides adopting regular, defined secondary structures. This approach has been reviewed by Krimm and Bandekar [14]. However, these methods cannot be extended to large proteins of irregular structure. Nevertheless, predicted frequencies for regular structures are in reasonable agreement with those observed for large irregular proteins of known structure.

Because theoretical methods are difficult to apply to large, irregular proteins, an empirical approach has been used by many workers to develop assignments that relate the positions of various amide band maxima to secondary structure. The amide I, II and III bands have proved most useful in this analysis. These bands are not caused by localized band stretching or bending motions but a detailed analysis of the delocalization of these in-plane vibrations is available [13, 15, 16]. The amide I band is principally (80%) a C=O stretching vibration with contributions from N—H bending and C—N stretching. The amide II band is principally N—H bending (60%) together with C—N stretching. The amide III is a more complex mode with major contributions from C—N stretching (30%) and N—H bending (30%), together with C=O stretching, O=C—N bending and other vibrations.

Theoretical [10, 14] and empirical approaches [11–13] indicate that the amide I band is most sensitive to secondary structure. Experimental evidence has been derived from studies of fibrous proteins, soluble proteins and homopolypeptides of known structure. In some cases it is possible to manipulate the secondary structure of a polypeptide in solution by variation of pH and/or temperature [11]. Because the amide I band overlaps with the H—O—H bending mode of water these studies were carried out either in deuterium oxide solution or using laborious methods to obtain the differential absorption of a protein in water. The use of deuterium oxide raises further complications owing to the small shift of the amide I band to lower wavenumber on isotopic exchange of amide N—H to N—^{2}H. However, by recording the band position in water and deuterium oxide the spectroscopist can obtain a more reliable assignment, as well as information on the solvent accessibility of secondary structure elements. Because most globular proteins contain several types of secondary structure, usually in short segments, there are bound to be discrepancies between observed amide I band positions and those calculated for regular, repeating and infinite structures [17–19] or those observed for homopolypeptides of known structure [11]. Nevertheless, these early assignments remain in broad agreement with the majority of the data now available on a wide range of structures.

The experimental problems of poor signal-to-noise, strong absorption of water and uncertainties arising from overlapping bands ensured that i.r. spectroscopy was not applied widely to the problems of protein structure until the 1980s. The commercial availability of a range of Fourier transform i.r. spectrometers in the late 1970s gave fresh impetus to this research. Fourier transform i.r. spectrometers have several advantages over the dispersive (grating) spectrometers which were used in the early studies. Because the frequencies are sampled simultaneously by the interferometer, many scans can be signal-averaged in a relatively short time with a dramatic improvement in signal-to-noise as a consequence (the Multiplex or Fellgett Advantage). The signal-to-noise is also improved by the increase in energy throughput derived from the absence of slits (Jacquinot Advantage). The use of a HeNe laser of precisely defined wavelength ensures accurate frequency calibration and reproducibility (Connes Advantage). In addition, unlike dispersive instruments, Fourier transform i.r. spectrometers provide constant resolution at all wavelengths and there is negligible stray light.

An early application of Fourier transform i.r. spectrometry [20] compared the solution and cast film spectra of several proteins whose structure is known from single-crystal X-ray diffraction. Spectra of proteins in water at 50 mg/ml were recorded at an extremely short pathlength (5–10 µm) to ensure that the water absorption did not attenuate the i.r. beam completely. A reference spectrum of water was digitally subtracted from the protein spectra to obtain difference spectra displaying relatively strong protein absorptions free from interference by solvent. This study concluded that the secondary structure of globular proteins can be distinguished by a combination of amide I band frequency and amide III band intensity. In particular, the amide I band for α-helix was found over the range 1652–1656 cm^{-1}, for β-sheet at 1632 cm^{-1} and for disordered structures at 1655 cm^{-1}. These positions are in accordance with the earlier studies described above [13]. They also concluded that the amide II band could not be used reliably for determination of secondary structure. The majority of studies performed since have concentrated on the amide I band, making assignments from the systematic studies on proteins of known structure carried out by Susi and co-workers [11–13, 21, 22].

Resolution enhancement of the i.r. spectra of proteins

The amide I band of globular proteins absorbs over the range 1610–1700 cm^{-1}. The position of the peak maximum reflects the major secondary structure present. As most globular proteins contain a variety of secondary structure elements it is clear from the above assignments that the minor structures may be concealed by such a broad feature. Least squares optimization [23], Fourier self-deconvolution [24] and derivative spectroscopy [25] have been developed for the observation of the individual components of overlapping bands. Least squares optimization is limited by its requirement for information on the number, likely positions and bandshapes of the overlapping components. Considerable advances have been made with the application of deconvolution and derivatives to the i.r. spectra of proteins. Although these methods have been termed resolution enhancement techniques, they do not increase the resolution of the spectrum, which is fixed by the instrumental parameters for data collection, but they do improve the visualization of overlapping components.

Details of the method for Fourier self-deconvolution are given elsewhere [24, 26]. Its early applications to the spectra of globular proteins revealed many of the component bands which were expected for complex structures [27, 28]. Other studies have used second-derivative spectra to reveal the same components [21, 29]. Both methods are essentially band-narrowing techniques and result in the observation of previously hidden components as distinct peaks with associated minima. Detailed reviews describing these methods and their applications to the spectra of proteins and membranes are available [30–32]. Although these techniques are very powerful, they must be applied with a careful appreciation of the likely artefacts.

Both derivative and deconvolved spectra exhibit a reduced signal-to-noise compared with the original data and it is therefore essential that the experimental conditions are fully optimized. A spectral region free from sample absorption should be compared with the intensity of the observed components to ensure that the spectrum is not over-interpreted. Deconvolved spectra may be distorted by the appearance of negative 'feet' either side of strong absorption bands. Even-order derivative spectra are distorted by similar features either side of each absorption which can in extreme cases lead to cancellation of a weaker component by the 'feet' of a neighbouring stronger band [33]. The amide I region of the spectrum is also complicated by the possibility of contributions by atmospheric water vapour. Since both deconvolved and derivative spectra emphasize sharp bands, any water-vapour absorption causes severe distortion of the spectra. It is therefore important that a strong purge with dry gas is used to eliminate these features [29].

As an illustration of the power of these techniques, Fig. 1 presents difference, second-derivative and deconvolved spectra of α-chymotrypsin in water. The difference spectrum, obtained by digital subtraction of a spectrum of water from that of 5% w/v solution of the protein, exhibits broad amide I and amide II bands at 1639 cm^{-1} and 1549 cm^{-1}, respectively. Both bands are shown to consist of a number of components in both the second-derivative and deconvolved spectra. The calculation of both types of resolution-enhanced spectra enables data to be interpreted with greater confidence for components that are reproducible. In particular, the band at 1636/7 cm^{-1} is consistent with the presence of β-structure, at least some of which is antiparallel as indicated by the presence of the band at 1687/9 cm^{-1}. The bands at 1648 cm^{-1} and 1658/60 cm^{-1} result from α-helix and α-helix and/or turn structures, respectively. It is apparent from these spectra that the second-derivative spectrum is most powerful in terms of separating overlapping bands. However, it is also clear that deconvolution results in less distortion of the band shapes. As a result it is possible to extract quantitative information from deconvolved spectra (see below) whereas derivative spectra are unreliable in this respect.

Quantitative analysis of the i.r. spectra of proteins

The application of these methods for the detailed analysis of the amide I and amide II bands in terms of the presence of types of secondary structure has been extended to quantitative methods. There are limitations to the use of resolution-enhanced spectra for quantitative analysis. Peak intensities in these spectra are dependent on the original bandwidths of the components which are usually unknown. Early quantitative methods employed curve-fitting directly to the amide I band [23, 34]. This approach requires the input of the number, positions and widths of all the component bands. More recently, approaches based on deconvolution and derivatives to identify the number and position of the components, followed by iterative curve-fitting to either the deconvolved [22, 35–37] or original difference spectra [38], have proved more successful. The areas of all fitted components assigned to a particular structure are summed and expressed as a percentage of the total amide I area.

Fig. 1. I.r. spectra of α-chymotrypsin from bovine pancreas in water at 20°C

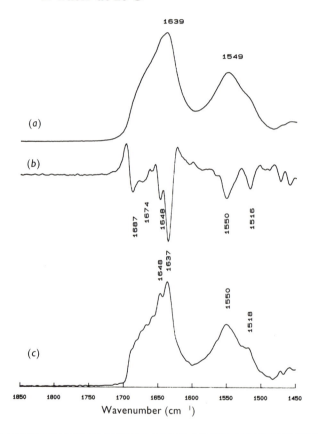

(a) difference spectrum; (b) second-derivative spectrum; (c) deconvolved spectrum. $\sigma = 6\ cm^{-1}$, $K = 2.0$.

Although these methods have shown remarkably good agreement between secondary structure as determined by i.r. spectrometry and the original X-ray data for a wide range of soluble proteins, a number of problems are associated with their application [39, 40]. As noted earlier, deconvolution can introduce spectral distortions and emphasize noise and water vapour. These methods also require assignment of each component of the amide I band.

There are many instances where it is difficult to make a complete assignment with certainty. In addition there are many difficulties encountered in curve-fitting, particularly in its reproducible application between laboratories. In many instances, in order to get a close fit, it is necessary to add components not identified in the deconvolved or derivative spectra. While the resolution-enhanced spectra may not reveal all the components, there is little justification in accepting these extra bands as an accurate reflection of structure. Not all of the absorption in the region of the amide I band may be caused by the protein backbone [39]. Therefore, although this approach has its limitations for comparative determinations, it remains a useful method for assessing the extent of conformational changes in a single system under different experimental conditions.

Alternatively, factor analysis [41] of the i.r. spectrum can provide quantitative assessments. Factor analysis is a whole spectrum approach which does not involve assignment of distinct spectral features to the quantified components. It has been applied successfully to the analysis of both soluble [40, 42, 43] and membrane [44] proteins. A detailed description of factor analysis methods is beyond the scope of this chapter and the interested reader is referred to the original papers [41, 45, 46]. Fig. 2 presents our comparison of the secondary structure of a series of proteins according to a partial least squares [46] method for factor analysis with the original X-ray data [47]. The agreement between the i.r. and X-ray values is good with standard errors of prediction of 4.8% for α-helix, 6.4% for β-sheet and 3.3% for turn structures [44]. While this method provides little opportunity for operator bias, it does require the generation of a calibration set consisting of the i.r. spectra of proteins of known structure.

Studies of the structure of membrane proteins

Bacteriorhodopsin
Bacteriorhodopsin is the single protein of the purple membrane of *Halobacterium halobium*. It utilizes light energy to pump protons across the membrane setting up an electrochemical gradient which is used by the cell to synthesize ATP. It comprises around 75% of the dry weight of the purple membrane, is easily purified, extremely stable and as a consequence is the most commonly used membrane protein for biophysical studies. In many respects it has become a model for understanding membrane transport processes. X-ray and electron diffraction studies of purple membrane films have shown that around 80% of the bacteriorhodopsin forms seven transmembrane α-helices tilted slightly with respect to the membrane normal [48, 49]. Recently, a combination of high-resolution electron microscopy and electron diffraction have provided a three-dimensional density map with resolutions of 0.35 nm and 1 nm, parallel and perpendicular to the membrane, respectively [50]. I.r. spectroscopy has played an important role in character-

Fig. 2. Secondary structure predictions by partial least squares analysis of the i.r. spectra (**IR-PLS**) of 17 soluble proteins compared with estimations [47] based on X-ray data (**XR**)

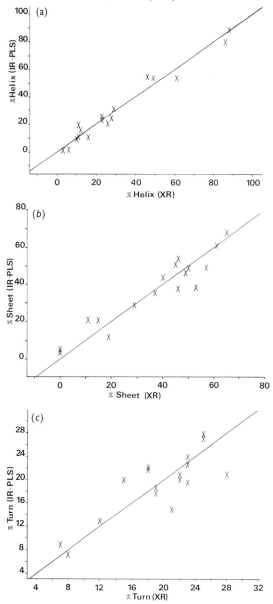

(a) α-helix; (b) β-sheets; (c) turns.

Fig. 3. Difference (*a*), second derivative (*b*) and deconvolved [$\sigma = 6.5$ cm^{-1}, $K = 2.0$] (*c*) spectra of bacteriorhodopsin in the purple membrane

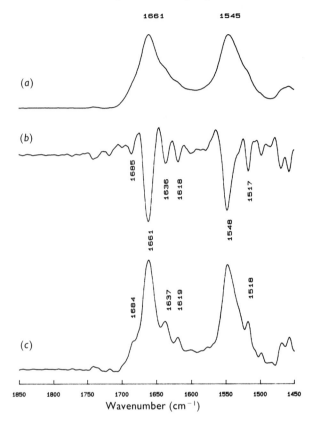

izing the secondary structure and orientation of transmembrane helices in the purple membrane, as well as the kinetics and involvement of specific amino acid residues in proton transport.

Early studies used polarized i.r. of purple membrane films [51] and difference [52] and derivative [29] analysis of purple membrane suspensions. Fig. 3 presents difference, second-derivative and deconvolved i.r. spectra of bacteriorhodopsin in the purple membrane. The presence of the major amide I band near 1660 cm^{-1} has been interpreted to result from 'distorted' α-helices. A comparison of the i.r. spectrum of purple membranes with normal mode calculations for poly(L-alanine) suggested that this anomalously high amide I band might be explained by the presence of α_{II}-helical structure [53]. Alpha$_{II}$ helices have the same number of residues per turn and rise per residue as α-helices but the plane of the peptide bond is tilted with the

N—H pointing toward the helical axis. This results in a lengthening and weakening of the C=O....H—N hydrogen bonds compared with the α-helix where the plane of the peptide bond is parallel to the helical axis.

The amide I band of bacteriorhodopsin also has a shoulder near 1635 cm^{-1} that is revealed as a separate component in the derivative and deconvolved spectra (Fig. 3) and has been assigned to the presence of β-structure [29, 52, 54]. An alternative assignment to incompletely subtracted water is contradicted by the presence of this component after extensive hydrogen–deuterium exchange [55]. A polarized i.r. study of purple membrane films indicated that any β-structure was not orientated perpendicular to the membrane [56]. Proteolytic digestion was used to show that the 1636 cm^{-1} component remained after removal of the extramembranous C-terminus, a loop between helices B and C and the N-terminus [55]. A further component at 1685 cm^{-1} has been assigned to β-sheet and/or turns.

In an attempt to resolve the discrepancies between the data from vibrational spectroscopy and those from diffraction studies, Haris and Chapman [57] compared i.r. spectra of alamethicin and bacteriorhodopsin. Alamethicin forms voltage-gated ion channels in membranes. Its structure has been studied extensively by X-ray diffraction of crystals from methanol solution and by n.m.r. spectroscopy of samples in water and methanol. The consensus of these studies is that in methanol and in the crystal, alamethicin is principally α-helical in structure. The crystal structure contains a proline bend with 3_{10}-helical structure at the C-terminus. The i.r. spectra of alamethicin in methanol solution and in phospholipid bilayers are very similar in the amide I region to the spectrum of bacteriorhodopsin in the purple membrane [57]. Since the crystal structure of alamethicin does not show either β-sheet or $α_{11}$-helical structures, the accepted i.r. band frequency correlations cannot be applied to alamethicin or, by analogy, to bacteriorhodopsin. Haris and Chapman suggest that the unusually high amide I frequency of bacteriorhodopsin arises from an overlap of the absorptions of both α-helices and 3_{10}-helices [57].

The orientation of the transmembrane helices of bacteriorhodopsin has been investigated by polarized i.r. spectroscopy. Studies on dehydrated [51, 56] and rehydrated [58] orientated purple membrane films are in close agreement and give an average tilt of 24–27 degrees. The latter study also examined hydrogen–deuterium exchange processes and concluded that while non-orientated secondary structures undergo rapid exchange the transmembrane helices are resistant to exchange even after 48 h of exposure to ^2H_2O. This is in agreement with earlier observations on purple membrane suspensions [55]. It was also shown that part of the 1640 cm^{-1} shoulder is caused by the C=N stretching vibration of the chromophore [58].

The photocycle of bacteriorhodopsin has been studied by i.r. difference spectroscopy. Intermediates in the photocycle may be stabilized at low temperatures and difference spectra calculated that correspond to the generation of the K, L or M intermediates [59, 60]. The interpretation of these difference spectra has been greatly assisted by the use of isotope labelling [61, 62] and site-directed mutagenesis [63, 64] to obtain assignments. From these results a model for the proton-pumping mechanism was proposed [64]. This described proton transfers between the retinal Schiff base and residues Asp-96, Asp-212 and Asp-85. This model is largely consistent with one developed from the latest structural data on bacteriorhodopsin [50].

Kinetic i.r. techniques permit studies of the photocyle of bacteriorhodopsin under ambient temperatures. The earliest approach involved monitoring absorption changes at fixed wavelengths [65]. More recent methods have exploited the increased sensitivity of Fourier transform spectrometers [66–70]. The kinetics of the deprotonation of Asp-96 and the protonation of Asp-85 are consistent with the proposed mechanism [64]. Structural changes in the protein during the photocycle have also been detected by kinetic [70] and low-temperature [71] techniques. Infrared studies of the bacteriorhodopsin photocycle have been extended recently to the picosecond time-scale [72].

Rhodopsin
Rhodopsin is the photoreceptor pigment of the rod cells of the vertebrate retina. Absorption of a photon triggers a change in the membrane potential and initiates a biochemical cascade involving cyclic GMP. The folding of rhodopsin in the lipid bilayer has been predicted from the amino acid sequence [73]. Rhodopsin is part of a receptor family that is functionally coupled to G proteins. This family includes the catecholamine receptors, the muscarinic acetylcholine receptors and many others. They are homologous proteins that contain seven transmembrane α-helices.

Early i.r. studies of rhodopsin in rod outer segment membranes concluded that it adopts a primarily α-helical conformation [74, 75]. More recent studies employing resolution enhancement techniques have supported a predominantly α-helical structure but have also revealed the presence of minor amide I components that have been assigned to β-sheet and β-turns [76, 77]. Fig. 4 shows a comparison of the deconvolved i.r. spectra of rhodopsin and bacteriorhodopsin. It is of interest that the main amide I component assigned to α-helices in the spectrum of unbleached rhodopsin is found at 1657 cm^{-1}, whereas the corresponding band in the spectrum of bacteriorhodopsin is found at 1661 cm^{-1} [77]. This implies that the details of the structure of bacteriorhodopsin may not provide an accurate model for G protein-coupled receptors. The α-helices of rhodopsin are orientated predominantly perpendicular to the membrane plane according to i.r. linear dichroism measurements [78].

The characterization of photointermediates and conformational changes during the photolytic cascade has been addressed by difference i.r. spectroscopy [79–81]. Photostationary states have been trapped at low

Fig. 4. Deconvolved i.r. spectra of bacteriorhodopsin (*a*) and rhodopsin (*b*).

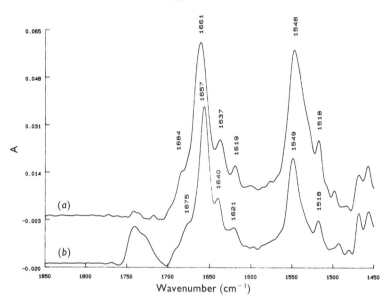

$\sigma = 6.5\ cm^{-1}$; $K = 2.0$.

temperatures and the reaction of rhodopsin to Meta II and Meta III examined at room temperature. These studies provide information on the protein–chromophore linkage, isomeric state and opsin conformation.

Photosynthetic membranes

There have been several detailed studies of the structure of proteins of the photosynthetic membranes of bacteria and green plants using i.r. spectroscopy [82–86]. The reaction centres of *Rhodobacter sphaeroides* and *Rhodopseudomonas viridis* gave amide I maxima of 1657–1658 cm^{-1} when examined as air-dried films indicating an α-helical conformation. Second-derivative, deconvolved and curve-fitted spectra have supported this assessment but also revealed the presence of β-sheet and turn structures [85].

A detailed i.r. study on the secondary structure of the photosystem II reaction centre included quantitative analysis and the effects of exposure to bright illumination [86]. The isolated photosystem II reaction centre contains a high proportion of α-helical structure together with some β-sheet and some turn structures according to the Fourier transform i.r. analysis. Quantitative estimations based on factor analysis indicate that the photosystem II reaction centre contains a higher proportion of α-helical structure than the bacterial reaction centres [86] (D. C. Lee, P. I. Haris, D. Chapman & R. C.

Mitchell, unpublished work). Exposure of the photosystem II reaction centre to bright white light causes a dramatic reduction in α-helical content and a corresponding increase in β-sheet.

The structures of the reaction centres of *Rp. viridis* and *Rb. sphaeroides* are known in detail from X-ray crystallography [3, 87]. These represent static pictures of quiescent states. A variety of i.r. difference techniques have provided functional information at the level of individual chemical groups. Light-induced difference spectroscopy has been used to probe molecular changes in P (the primary electron donor — a dimer of bacteriochlorophyll) and Q_A (the primary acceptor — a quinone) on charge separation. Molecular changes in both P and Q_A together with their protein environment may be investigated on photogeneration of $P^+Q_A^-$ [88–90]. The contributions of P and P^+ dominate the $P^+Q_A^-/PQ_A$ difference spectrum but the models affected by reduction of Q_A may be probed by blocking the electron transfer to the secondary acceptor and reducing P^+ [91]. The P/P^+ and Q_A/Q_A^- transitions may also be isolated electrochemically and Fourier transform i.r. difference spectra examined [89, 92, 93]. Polarized i.r. techniques have been applied to investigations of the orientation, relative to the membrane plane, of the vibrational modes affected by charge separation [94].

Ca^{2+}-ATPase

The Ca^{2+}-transporting ATPase represents the major protein of the sarcoplasmic reticulum of skeletal muscle and is one of the best characterized of ion-transporting proteins. The sarcoplasmic reticulum is a membrane system involved in the regulation of the contraction–relaxation cycle of striated muscle. In response to neuronal stimulation of the muscle, calcium ions are rapidly released from the sarcoplasmic reticulum to bind to troponin and initiate contraction. Cessation of stimulation results in Ca^{2+} accumulation by the sarcoplasmic reticulum such that the intracellular concentration falls below 10^{-7} M and muscle relaxation occurs. The active accumulation of Ca^{2+} during relaxation is carried out by the Ca^{2+}-ATPase using energy derived from the hydrolysis of ATP.

The primary sequence of the Ca^{2+}-ATPase is known from a cDNA clone and models for the secondary structure have been proposed [95, 96]. An early difference i.r. study of sarcoplasmic reticulum revealed a predominantly α-helical conformation [52]. Subsequent studies employed deconvolution [97] and derivative [29, 98] techniques to observe overlapping components of the amide I band and revealed the presence of a small amount of β-sheet structure. The generation of a fourth-derivative spectrum of good signal-to-noise permitted the identification of a band caused by turns [98]. This qualitative assessment is in agreement with the secondary structure predicted by analysis on the primary sequence [95, 96]. More recent studies have sought a quantitative description. Villalain *et al.* [99] performed a band-fitting analysis to their difference spectra of purified Ca^{2+}-ATPase and found 45% α-helix, 32% β-sheet and 23% turn structure. This analysis was obtained by summing the areas of the components assigned to each structure. The fact that no component could be assigned to random structure illustrates an inherent weakness in this approach. Our more recent study, employing a partial least squares analysis, gives a composition of 65% α-helix, 16% β-sheet and 7% turn structures [44]. This is in reasonable agreement with the secondary structure prediction based on the sequence, which indicates 52% α-helix and 12% β-sheet [95]. The prediction envisages 10 transmembrane α-helices linked to a cytoplasmic domain by an α-helical stalk region. The nucleotide-binding and phosphorylation domains are found in regions of mixed α-helix/β-sheet structure.

I.r. spectroscopy has been applied in attempts to elucidate the structural changes associated with the function of the Ca^{2+}-ATPase. A cyclic reaction scheme was proposed which described calcium binding and phosphorylation by ATP and involved two major conformations E_1 and E_2 with calcium-binding sites on opposite sides of the membrane [100]. There have been many attempts to provide biophysical evidence for this two-state model. Although there is an increase in intrinsic tryptophan fluorescence and an alteration in the electron spin resonance spectrum on Ca^{2+} binding, there is no alteration in the circular dichroism spectrum. There have also been conflicting reports from studies using i.r. spectrometry. Arrondo et al. [101], studying isolated sarcoplasmic reticulum vesicles, described the appearance of a small band assigned to α-helices on conversion from E_1 (Ca^{2+}-bound) to E_2 (Ca^{2+}-free). However, Villalain et al. [99] could not reproduce this result with reconstituted Ca^{2+}-ATPase. This discrepancy may be due to the difference in the preparation of the protein and/or the difficulties in obtaining reproducible spectra after the addition of ligands. The small spectral changes observed [101] may not be significant compared with uncertainties in buffer subtraction and alterations in protein concentration.

More precise measurements have been obtained using the techniques for difference spectroscopy (between altered states) applied previously to the photogenerated forms of bacteriorhodopsin and rhodopsin. In order to apply these techniques to the Ca^{2+}-ATPase and other light-insensitive systems, methods must be devised for triggering and controlling the appropriate reaction. An elegant solution is the use of ligands trapped in a photolabile cage, thereby enabling the reaction to be initiated without replacing the sample in the i.r. cell. Barth et al. [102] used caged ATP to trigger phosphorylation of the Ca^{2+}-ATPase. The changes in the i.r. spectra were small compared with the total absorbance, indicating that phosphorylation is not accompanied by a major conformational rearrangement. Instead, the difference bands were assigned to localized structural modifications near the phosphorylation and calcium-binding sites. Similarly, Buchet et al. [103] used caged Ca^{2+} to effect the formation of the proposed Ca_2E_1 intermediate in the absence of ATP. Fig. 5 presents difference i.r. spectra of sarcoplasmic reticulum in the presence of caged Ca^{2+} after minus before release of Ca^{2+} by photolysis of the cage [103]. The spectral changes are small but highly

Fig. 5. Difference i.r. spectra of sarcoplasmic reticulum after minus before release of Ca^{2+} from a Nitr-5–Ca^{2+} complex

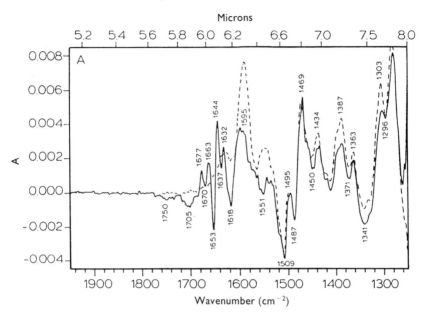

The spectra (solid lines) represent the averages of six difference spectra (after minus before). As a control the same experiment was performed in the absence of sarcoplasmic reticulum (broken lines). Figure reproduced from [103], with permission.

reproducible. The features at 1677 cm^{-1}, 1663 cm^{-1} and 1653 cm^{-1} can be explained by Ca^{2+}-induced changes in the sarcoplasmic reticulum proteins, as they are not observed on photolysis of the cage in the absence of protein (broken line). The feature at 1644 cm^{-1} was eliminated by addition of dithiothreitol to prevent the reaction of cage photoproducts with the sarcoplasmic reticulum. The authors favoured an interpretation involving alterations in ionization of amino acid side-chains rather than changes in secondary structure.

Other ATPases

I.r. spectroscopy has also been used to study the structure and conformational changes in other membrane-bound ATPases. The gastric H^+/K^+-ATPase is responsible for the secretion of acid by the parietal cell and represents an attractive target for pharmacological intervention. Mitchell *et al.* [104], using i.r. spectroscopy, showed that the gastric H^+/K^+-ATPase contains substantial amounts of both α-helical and β-sheet structure. They also investigated conformational changes associated with function and the binding of an inhibitor. Addition of K^+ to produce the E_2K^+ form resulted in no change in the amide I in the second-derivative spectrum suggesting no gross alteration in secondary structure. The addition of omeprazole, a potent inhibitor of gastric acid secretion which acts by binding irreversibly to thiol groups on the H^+/K^+-ATPase, did induce significant conformational change. Fig. 6 shows second-derivative i.r. spectra of the gastric H^+/K^+-ATPase in the presence of omeprazole recorded 21 min and 365 min after the initiation of $^1H-^2H$ exchange. The decrease in the amide II band at 1547 cm^{-1} was much greater than that observed in the absence of omeprazole and suggests that the inhibitor increases the solvent accessibility of the protein. The band at 1627 cm^{-1} is not observed in the absence of omeprazole and, unlike the α-helix band at 1655 cm^{-1}, the band assigned to β-sheet at 1634 cm^{-1} has shifted significantly on isotopic exchange. This suggests that binding of omeprazole may cause perturbation of β-sheet structures.

Fig. 6. Second-derivative i.r. spectra of omeprazole-inhibited gastric H^+/K^+-ATPase in ^2H_2O recorded 21 min (solid line) and 365 min (broken line) after the addition of ^2H_2O

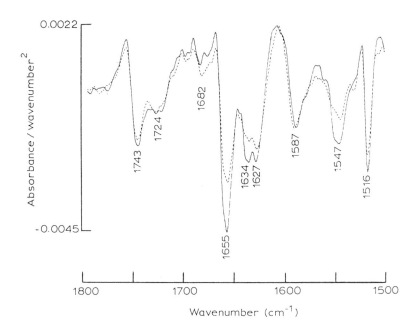

Figure reproduced from [104], with permission.

There have been several i.r. studies on the conformation of the Na^+/K^+-ATPase from pig renal medulla [104-108]. The early studies were interpreted to suggest that this protein contained 55% unordered structure and 20-25% each of α-helix and β-sheet. These experiments were performed using dispersive i.r. spectrometers and rather primitive techniques for the observation of overlapping components of the amide I band. More recently, a Fourier transform i.r. study could not detect any amide I component assignable to unordered structure, suggesting that 55% is a gross overestimate of its contribution [104]. Evidence has also been sought for conformational changes on conversion between the Na^+ and K^+-bound forms of the enzyme [106, 107]. No significant (greater than 2-3%) differences were found. A more recent study by Raman spectroscopy [108] contradicts these findings and concludes that approximately 100 peptide groups are converted from β-sheet to α-helical conformation by the E_1Na^+-E_2K^+ transition. These authors suggest that the absence of any changes in the i.r. experiments is the result of enzyme inhibition by the ^2H_2O media. The structure of the Na^+/K^+-ATPase has also been studied by polarized attenuated total reflection i.r. spectroscopy [109]. Here it was suggested that approximately 13% of the amino acids were involved in a structural transition on exposure to K^+. Changes in the absorption of the carboxylate modes of glutamate and aspartate were also detected.

Human erythrocyte glucose transporter
The transport of glucose across the membrane of the human erythrocyte occurs by facilitated diffusion mediated by a protein of M_r 55 000. This protein has been purified and reconstituted and its transport kinetics and inhibition studied in some detail. Several structural studies by i.r. spectroscopy have been carried out. Second-derivative spectra of the purified protein reconstituted into erythrocyte lipids reveal a predominantly α-helical structure with some β-structure [110]. A polarized i.r. study also described a largely α-helical protein with an effective tilt of the helices of less than 38° from the membrane normal [111]. This tilt angle was reduced in the presence of the substrate D-glucose. However, this study did not detect any β-structure, either by curve-fitting to the amide I band or by examination of the amide A band. The same group studied the hydrogen-deuterium exchange behaviour using i.r. spectroscopy together with hydrogen-tritium exchange and concluded that a large proportion of the amide hydrogens in the transmembrane regions are accessible to water [112]. It was concluded that this is consistent with the presence of an aqueous transmembrane channel. The kinetics of exchange were perturbed by D-glucose and the inhibitor cytochalasin B.

Alvarez et al. [113] used second- and fourth-derivative and deconvolved spectra in the amide I region to analyse the secondary structure, conformational changes and the time-dependency of hydrogen-deuterium exchange. A principally α-helical structure was confirmed, in agreement with a model predicted from the amino acid sequence [114]. The resolution enhanced spectra revealed a relatively weak component, near 1630 cm^{-1}, assigned to β-sheet. It was shown that 75% of the amide hydrogens undergo deuterium exchange within the first hour of transfer of the protein to a ^{2}H$_2$O buffer. This accessibility to exchange contrasts with the behaviour of other membrane proteins and is consistent with the presence of an aqueous channel in the protein. The protein was incubated with substrate and substrate analogues in order to investigate the proposed alternating conformation model for transport. Both incubation with ethylidene-D-glucose and increased temperature, which largely fix the transporter in the 'outward-facing' conformation, gave a small but reproducible shift in the α-helix band compared with conditions in which the 'inward-facing' form predominates.

Other membrane proteins

The systems described above represent those membrane proteins for which most detailed i.r. studies are currently available. However, important studies have also been carried out on porin [115, 116], cytochrome oxidase [117], the proteins of myelin membranes [118–120] and cytochrome b_5 [121]. Porin is unusual among integral membrane proteins in consisting predominantly of anti-parallel β-sheet. Infrared linear dichroism measurements indicate that the average orientation of the β-sheets is approximately normal to the membrane plane [116]. Lipophilin, the hydrophobic protein of myelin, is predominantly α-helical with a significant amount of β-sheet. The proportion of β-sheet is increased with increasing temperature [120]. Myelin basic protein is a peripheral membrane protein that adopts a largely unfolded conformation when prepared in aqueous solution [119]. However, interaction of myelin based protein with the acidic phospholipid, phosphatidylglycerol, resulted in large changes in the amide I band, indicative of significant ordering to β-sheet [119].

> *My thanks go to Dennis Chapman, Parvez Haris and Bob Mitchell for many stimulating discussions on the application of i.r. spectroscopy in structural studies of proteins, lipids and membranes.*

References

1. Michel, H. (1983) Trends Biochem. Sci. 8, 56–59
2. Deisenhofer, J., Epp, O., Miki, M., Huber, R. & Michel, H. (1984) J. Mol. Biol. 180, 385–395
3. Allen, J. P., Feher, G., Yeates, T. O., Komiya, H. & Rees, D. C. (1987) Proc. Natl. Acad. Sci. U.S.A. 84, 6162–6166
4. Weiss, M. S., Wacker, T., Weckesser, J., Welte, W. & Schulz, G. E. (1990) FEBS Lett. 267, 268–272
5. Kyte, J. & Doolittle, R. (1982) J. Mol. Biol. 157, 105–132
6. Engelman, D. M., Steitz, T. A. & Goldman, A. (1986) Annu. Rev. Biophys. Chem. 15, 321–353
7. Chou, P. Y. & Fasman, G. D. (1974) Biochemistry 13, 222–244
8. Wallace, B. A., Cascio, M. & Mielke, D. L. (1986) Proc. Natl. Acad. Sci. U.S.A. 83, 9423–9427
9. Elliot, A. & Ambrose, E. J. (1950) Nature (London) 165, 921–922
10. Krimm, S. (1962) J. Mol. Biol. 4, 528–540

11. Susi, H., Timasheff, S. N. & Stevens, L. (1967) J. Biol. Chem. 242, 5460–5466
12. Timasheff, S. N., Susi, H. & Stevens, L. (1967) J. Biol. Chem. 242, 5467–5473
13. Susi, H. (1969) in Structure and Stability of Biological Macromolecules (Timasheff, S. N. & Fasman, G. D., eds.), pp. 575–663, Decker, New York
14. Krimm, S. & Bandekar, J. (1986) Adv. Protein Chem. 38, 181–364
15. Miyazawa, T., Shimanouchi, T. & Mizushima, S. (1956) J. Chem. Phys. 24, 408–418
16. Miyazawa, T., Shimanouchi, T. & Mizushima, S. (1958) J. Chem. Phys. 29, 611–616
17. Chirgadze, Y. N. & Nevskaya, N. A. (1976) Biopolymers 15, 607–625
18. Chirgadze, Y. N. & Nevskaya, N. A. (1976) Biopolymers 15, 627–636
19. Nevskaya, N. A. & Chirgadze, Y. N. (1976) Biopolymers 15, 637–648
20. Koenig, J. L. & Tabb, D. L. (1980) in Analytical Applications of FT-IR to Molecular and Biological Systems (Durig, J. R., ed.), pp. 241–255, Reidel, Holland
21. Susi, H. & Byler, D. M. (1983) Biochem. Biophys. Res. Commun. 115, 391–397
22. Byler, D. M. & Susi, H. (1986) Biopolymers 25, 469–487
23. Ruegg, M., Metzger, V. & Susi, H. (1975) Biopolymers 14, 1465–1471
24. Kauppinen, J. K., Moffat, D. J., Mantsch, H. H. & Cameron, D. G. (1981) Appl. Spectrosc. 35, 271–277
25. Maddams, W. F. & Southon, M. J. (1982) Spectrochim. Acta 38A, 459–466
26. Cameron, D. G. & Moffat, D. J. (1984) J. Testing Eval. 12, 78–85
27. Dev, S. B., Rha, C. & Walder, F. (1984) J. Biomolec. Struct. Dynamics 2, 431–442
28. Yang, W.-J., Griffiths, P. R., Byler, D. M. & Susi, H. (1985) Appl. Spectrosc. 39, 282–287
29. Lee, D. C., Hayward, J. A., Restall, C. J. & Chapman, D. (1985) Biochemistry 24, 4364–4373
30. Mantsch, H. H., Casal, H. L. & Jones, R. N. (1986) in Spectroscopy of Biological Systems (Clark, R. J. H. & Hester, R. E., eds.), pp. 1–46, Wiley, Chichester
31. Surewicz, W. K. & Mantsch, H. H. (1988) Biochim. Biophys. Acta 952, 115–130
32. Susi, H. & Byler, D. M. (1988) in Methods for Protein Analysis (Cherry, J. P. & Barford, R. A., eds.), pp. 235–255, American Oil Chemists' Society, Champaign, IL
33. Maddams, W. F. & Tooke, P. B. (1982) J. Macromol. Sci. Chem. A17, 951–968
34. Eckert, M., Grosse, R., Malur, J. & Repke, K. R. M. (1977) Biopolymers 16, 2549–2563
35. Susi, H., Byler, D. M. & Purcell, J. M. (1985) J. Biochem. Biophys. Methods 11, 235–240
36. Surewicz, W. K., Moscarello, M. A. & Mantsch, H. H. (1987) J. Biol. Chem. 262, 8598–8602
37. Holloway, P. W. & Mantsch, H. H. (1989) Biochemistry 28, 931–935
38. Arrondo, J. L. R., Young, N. M. & Mantsch, H. H. (1988) Biochim. Biophys. Acta 952, 261–268
39. Mantsch, H. H., Surewicz, W. K., Muga, A., Moffat, D. J. & Casal, H. L. (1989) Proc. SPIE Int. Soc. Opt. Eng. 1145, 580–581
40. Lee, D. C., Haris, P. I., Chapman, D. & Mitchell, R. C. (1990) Biochemistry 29, 9185–9193
41. Malinowski, E. R. (1991) Factor Analysis in Chemistry, 2nd edition, Wiley, Chichester
42. Lee, D. C., Haris, P. I., Mitchell, R. C. & Chapman, D. (1989) in Spectroscopy of Biological Molecules — State of the Art (Bertoluzza, A., Fagnano, C. & Monti, P. eds.), pp. 57–58, Societa Editrice Esculapio, Bologna, Italy
43. Dousseau, F. & Pezolet, M. (1990) Biochemistry 29, 8771–8779
44. Lee, D. C., Haris, P. I., Chapman, D. & Mitchell, R. C. (1991) in Spectroscopy of Biological Molecules (Hester, R. E. & Girling, R. B., eds.), pp. 7–10, Royal Society of Chemistry, Cambridge
45. Fredericks, P. M., Lee, J. R., Osborn, P. R. & Swinkels, D. A. J. (1985) Appl. Spectrosc. 39, 303–310
46. Haaland, D. M. & Thomas, E. V. (1988) Anal. Chem. 60, 1193–1202
47. Levitt, M. & Greer, J. (1977) J. Mol. Biol. 114, 181–293
48. Unwin, P. N. T. & Henderson, R. (1975) J. Mol. Biol. 94, 425–440
49. Henderson, R. & Unwin, P. N. T. (1975) Nature (London) 257, 28–32
50. Henderson, R., Baldwin, J. M., Ceska, T. A., Zemlin, F., Beckman, E. & Downing, K. H. (1990) J. Mol. Biol. 213, 899–929
51. Rothschild, K. J. & Clark, N. A. (1979) Biophys. J. 25, 473–487
52. Cortijo, M., Alonso, A., Gomez-Fernandez, J. C. & Chapman, D. (1982) J. Mol. Biol. 157, 597–618
53. Krimm, S. & Dwivedi, A. M. (1982) Science 216, 407–408
54. Jap, B. K., Maestre, M. F., Hayward, S. G. & Glaeser, R. M. (1983) Biophys. J. 43, 81–89
55. Lee, D. C., Herzyk, E. & Chapman, D. (1987) Biochemistry 26, 5775–5783
56. Nabedryk, E., Bardin, A. M. & Breton, J. (1985) Biophys. J. 48, 873–876
57. Haris, P. I. & Chapman, D. (1988) Biochim. Biophys. Acta 943, 375–380
58. Earnest, T. N., Herzfeld, J. & Rothschild, K. J. (1990) Biophys. J. 58, 1539–1546

59. Rothschild, K. J. & Marrero, H. (1982) Proc. Natl. Acad. Sci. U.S.A. **79**, 4045-4049
60. Bagley, K., Dollinger, G., Eisenstein, L., Singh, A. K. & Zimany, L. (1982) Proc. Natl. Acad. Sci. U.S.A. **79**, 4972-4976
61. Engelhard, M., Gerwert, K., Hess, B., Kreutz, W. & Siebert, F. (1985) Biochemistry **24**, 400-407
62. Roepe, P., Ahl, P. L., Das Gupta, S. K., Herzfeld, J. & Rothschild, K. J. (1987) Biochemistry **26**, 6696-6706
63. Braiman, M. S., Mogi, T., Stern, L. J., Hackett, N. R., Chao, B. H., Khorana, H. G. & Rothschild, K. J. (1988) Proteins: Struct. Funct. Genet. **3**, 219-229
64. Braiman, M. S., Mogi, T., Marti, T., Stern, L. J., Khorana, H. G. & Rothschild, K. J. (1988) Biochemistry **27**, 8516-8520
65. Siebert, F. & Mantele, W. (1980) Biophys. Struct. Mech. **6**, 147-164
66. Dollinger, G. D., Eisenstein, L. J., Croteau, A. A. & Alben, J. O. (1985) Biophys. J. **47**, 99a
67. Braiman, M. S., Ahl, P. L. & Rothschild, K. J. (1985) in Spectroscopy of Biological Molecules (Alix, A. J. P., Bernard, L. & Manfait, M., eds.), pp. 57-59, Wiley, Chichester
68. Braiman, M. S., Ahl, P. L. & Rothschild, K. J. (1987) Proc. Natl. Acad. Sci. U.S.A. **84**, 5221-5225
69. Gerwert, K., Souvignier, G. & Hess, B. (1990) Proc. Natl. Acad. Sci. U.S.A. **87**, 9774-9778
70. Braiman, M. S., Bousche, O. & Rothschild, K. J. (1991) Proc. Natl. Acad. Sci. U.S.A. **88**, 2388-2392
71. Ormos, P. (1991) Proc. Natl. Acad. Sci. U.S.A. **88**, 473-477
72. Diller, R., Iannone, M., Bogomolni, R. & Hochstrasser, R. M. (1991) Biophys. J. **60**, 286-289
73. Findlay, J. B. C. & Papin, D. J. (1986) Biochem. J. **238**, 625-642
74. Osborne, H. B. & Nabedryk-Viala, E. (1977) FEBS Lett. **84**, 217-220
75. Rothschild, K. J., DeGrip, W. J. & Sanches, R. (1980) Biochim. Biophys. Acta **596**, 338-351
76. Downer, N. W., Bruchman, T. J. & Hazard, J. H. (1986) J. Biol. Chem. **261**, 3640-3647
77. Haris, P. I., Coke, M. & Chapman, D. (1989) Biochim. Biophys. Acta **995**, 160-167
78. Rothschild, K. J., Sanches, R., Hsiao, T. L. & Clark, N. A. (1980) Biophys. J. **31**, 53-64
79. Bagley, K. A., Balogh-Nair, V., Croteau, A. A., Dollinger, G., Ebrey, T. G., Eisenstein, L., Hong, M. K., Nakanishi, K. & Vittitow, J. (1985) Biochemistry **24**, 6055-6071
80. Rothschild, K. J., Gillespie, J. & De Grip, W. J. (1987) Biophys. J. **51**, 345-350
81. De Grip, W. J., Gray, D., Gillespie, J., Bovee, P. H. M., Van Den Berg, E. M. M., Lugtenberg, J. & Rothschild, K. J. (1988) Photochem. Photobiol. **48**, 497-504
82. Nabedryk, E. & Breton, J. (1981) Biochim. Biophys. Acta **635**, 515-524
83. Nabedryk, E., Tiede, D. M., Dutton, P. L. & Breton, J. (1982) Biochim. Biophys. Acta **682**, 273-280
84. Nabedryk, E., Berger, G., Adrianambinintsoa, S. & Breton, J. (1985) Biochim. Biophys. Acta **809**, 271-276
85. Nabedryk, E., Breton, J. & Thibodeau, D. L. (1991) in Spectroscopy of Biological Molecules (Hester, R. E. & Girling, R. B., eds), pp. 67-68, Royal Society of Chemistry, Cambridge
86. He, W-Z., Newell, W. R., Haris, P. I., Chapman, D. & Barber, J. (1991) Biochemistry **30**, 4552-4559
87. Deisenhofer, J., Epp, O., Miki, M., Huber, R. & Michel, H. (1985) Nature (London) **318**, 618-624
88. Mantele, W., Nabedryk, E., Tavitian, B. A., Kreutz, W. & Breton, J. (1985) FEBS Lett. **187**, 227-232
89. Mantele, W., Wollenweber, A. M., Nabedryk, E. & Breton, J. (1988) Proc. Natl. Acad. Sci. U.S.A. **85**, 8468-8472
90. Nabedryk, E., Bagley, K. A., Thibodeau, D. L., Bauscher, M., Mantele, W. & Breton, J. (1990) FEBS Lett. **266**, 59-62
91. Breton, J., Thibodeau, D. L., Berthomieu, C., Mantele, W., Vermeglio, A. & Nabedryk, E. (1991) FEBS Lett. **278**, 257-260
92. Leonhard, M., Moss, D., Bauscher, M., Nabedryk, E., Breton, J. & Mantele, W. (1991) in Spectroscopy of Biological Molecules (Hester, R. E. & Girling, R. B., eds.), pp. 75-76, Royal Society of Chemistry, Cambridge
93. Bauscher, M., Moss, D. A., Leonhard, M., Nabedryk, E., Breton, J. & Mantele, W. (1991) in Spectroscopy of Biological Molecules (Hester, R. E. & Girling, R. B., eds.), pp. 77-78, Royal Society of Chemistry, Cambridge
94. Thibodeau, D. L., Nabedryk, E. & Breton, J. (1991) in Spectroscopy of Biological Molecules (Hester, R. E. & Girling, R. B., eds.), pp. 69-70, Royal Society of Chemistry, Cambridge
95. MacLennan, D. H., Brandl, C. J., Korczak, B. & Green, N. M. (1985) Nature (London) **316**, 696-700
96. Brandl, C. J., Green, N. M., Korczak, B. & MacLennan, D. H. (1986) Cell **44**, 597-607

97. Mendelsohn, R., Anderle, G., Jaworsky, M., Mantsch, H. H. & Dluhy, R. A. (1984) Biochim. Biophys. Acta 775, 215–224
98. Lee, D. C. & Chapman, D. (1986) Biosci. Rep. 6, 235–256
99. Villalain, J., Gomez-Fernandez, J. C., Jackson, M. & Chapman, D. (1989) Biochim. Biophys. Acta 978, 305–312
100. De Meis, L. & Vianna, A. (1979) Ann. Rev. Biochem. 48, 275–292
101. Arrondo, J. L. R., Mantsch, H. H., Mullner, N., Pikula, S. & Martonosi, A. (1987) J. Biol. Chem. 262, 9037–9043
102. Barth, A., Mantele, W. & Kreutz, W. (1991) Biochim. Biophys. Acta 1057, 115–123
103. Buchet, R., Jona, I. & Martonosi, A. (1991) Biochim. Biophys. Acta 1069, 209–217
104. Mitchell, R. C., Haris, P. I., Fallowfield, C., Keeling, D. J. & Chapman, D. (1988) Biochim. Biophys. Acta 941, 31–38
105. Brazhnikov, E. V., Chertverin, A. B. & Chirgadze, Yu, N. (1978) FEBS Lett. 93, 125–128
106. Chetverin, A. B., Venyaminov, S. Yu. N., Emelyanenko, V. I. & Burstein, E. A. (1980) Eur. J. Biochem. 108, 149–156
107. Chetverin, A. B. & Brazhnikov, E. V. (1985) J. Biol. Chem. 260, 7817–7819
108. Nabiev, I. R., Dzhandzhugazyan, K. N., Efremov, R. G. & Modyanov, N. N. (1988) FEBS Lett. 236, 235–239
109. Fringeli, U. P., Apell, H. -J., Fringeli, M. & Lauger, P. (1988) Biochim. Biophys. Acta 984, 301–312
110. Lee, D. C., Elliot, D. A., Baldwin, S. A. & Chapman, D. (1985) Biochem. Soc. Trans. 13, 684–685
111. Chin, J. J., Jung, E. K. Y. & Jung, C. Y. (1986) J. Biol. Chem. 261, 7101–7104
112. Jung, E. K. Y., Chin, J. J. & Jung, C. Y. (1986) J. Biol. Chem. 261, 9155–9160
113. Alvarez, J., Lee, D. C., Baldwin, S. A. & Chapman, D. (1987) J. Biol. Chem. 262, 3502–3509
114. Mueckler, M., Caruso, C., Baldwin, S. A., Panico, M., Blench, I., Morris, H. R., Leinhard, G. E., Allard, W. J. & Lodish, H. F. (1985) Science 229, 941–945
115. Kleffel, B., Garavito, R. M., Baumeister, W. & Rosenbusch, J. P. (1985) EMBO J. 4, 1589–1592
116. Nabedryk, E., Garavito, R. M. & Breton, J. (1988) Biophys. J. 53, 671–676
117. Bazzi, M. D. & Woody, R. W. (1985) Biophys. J. 48, 957–966
118. Surewicz, W. K., Moscarello, M. A. & Mantsch, H. H. (1987) J. Biol. Chem. 262, 8598–8602
119. Surewicz, W. K., Moscarello, M. A. & Mantsch, H. H. (1987) Biochemistry 26, 3881–3886
120. Carmona, P., de Cozar, M., Garcia-Segura, L. M. & Monreal, J. (1988) Eur. Biophys. J. 16, 169–176
121. Holloway, P. W. & Mantsch, H. H. (1990) Biochemistry 29, 9631–9637

Biocompatible surfaces based upon biomembrane mimicry

Y. P. Yianni

Biocompatibles Ltd, Brunel Science Park, Kingston Lane, Uxbridge, Middlesex UB8 3PQ, U.K.

Introduction

A wide range of materials and polymers are used in medical device applications. These include polyethylene, polypropylene, polyvinylchloride (PVC), polyesters, polystyrene, polyurethane, silicone, polysulphone, polyamide, polytetrafluoroethylene and cellulose and its derivatives. Although these have excellent mechanical and physical properties they were originally developed for use in industrial manufacturing and not specifically for the biomedical field.

The materials generally used for the fabrication of devices for biomedical application elicit adverse reactions when in contact with body tissues and fluids and especially blood. Contact of a foreign material with blood leads to protein deposition, platelet adhesion and activation, activation of the clotting cascade and, finally, production of a potentially fatal thrombus [1]. An additional systemic effect which may lead to clinically significant events is activation of the complement cascade. This is particularly a problem in procedures such as haemodialysis which involve blood contact with a very large surface area of a medical device [2].

To overcome these effects procedures such as cardiopulmonary bypass can only be undertaken if accompanied by systemic administration of heparin which acts as anticoagulant. This can, however, lead to complications such as uncontrolled bleeding and increased transfusion requirements (with accompanying risk of viral infection) [3]. Neutralization of heparin is required following the treatment. Although clotting is inhibited by use of heparin, many of the initial events leading to thrombus formation do actually proceed unhindered. Protein adsorption, platelet adhesion and activation at the material–blood interface and platelet aggregation within the blood itself therefore occur and these events all contribute to a dramatic lowering of the quality of a patient's blood during such extracorporeal procedures.

Some biomedical applications are totally precluded by the thrombogenic potential of synthetic polymers. An example of this is the small diameter vascular graft for use in coronary artery bypass. At present, all synthetic materials fail in this application and a patient requiring coronary bypass must first undergo a procedure to remove the saphenous vein from the leg. This vein is subsequently used to carry out the bypass itself. A biocompatible material would have an enormous benefit in this application.

The risk of thrombus formation is, in the case of some devices, also accompanied by the additional risk of infection as a result of the adhesion and proliferation of bacteria at the surface of a biomaterial. This is particularly evident in the use of in-dwelling catheters and is a direct consequence of the surface properties of the material used — usually polyurethane or silicone.

Blood is not the only body fluid that reacts adversely when in contact with a synthetic polymer. Hydrogel contact lenses for example suffer from the problem of protein deposition. This can cause considerable discomfort to the wearer and may in some cases preclude their use. In addition, urinary catheters suffer from mineral encrustation, which results in blockage that necessitates frequent replacement of the device.

Because of these problems associated with the use of current biomedical materials and the increasing clinical use of devices, there is a considerable requirement for improved biomaterials that do not illicit adverse reactions when in contact with body fluids and tissues. The issue of biocompatibility, that is, the property of interfacing with a biological system without modifying or adversely affecting its normal function, is therefore becoming ever more important within the medical device industry. The clinical application of devices with improved biocompatibility will therefore have an enormous impact in the medical field and will result in considerable improvements in patient safety and care, with a resultant enhancement of the quality of patients' lives.

Phosphorylcholine and biocompatibility

In recent years, the many problems and limitations associated with the use of synthetic polymers in medicine have lead to intensive research into the development of new materials with a reduced propensity for eliciting adverse biological responses such as thrombosis. Such materials would be termed 'biocompatible'.

One approach to the production of biocompatible surfaces was proposed by Dennis Chapman in the late 1970s and is based upon the modification of surface properties of materials using phosphorylcholine (PC) to produce interfacial characteristics which largely mimic the main lipid headgroup component of the outer surface of a natural cell membrane [4]. The feature required to achieve this, the PC headgroup, is found in biomembranes in the form of glycerophospholipid and sphingomyelin. This approach to biocompatibility exploits the inherently low levels of interaction between proteins and cells with PC, especially when it is presented as a close-packed, ordered, planar array at an interface [5, 6].

Fig. 1. A representation of a biological membrane showing the basic structural unit made up from a lipid bilayer containing proteins and glycoproteins

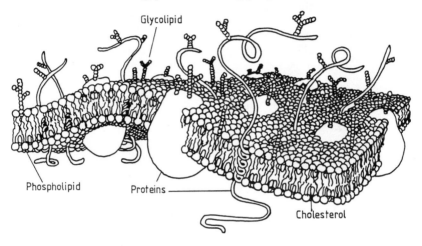

(Courtesy of B. Hall)

It was Zwaal *et al.* [7] who demonstrated the difference in thrombogenic potential of the two sides of a membrane, using right-side-out and non-resealed erythrocyte ghosts. The latter were shown to reduce the clotting time of blood in a concentration-dependent manner, whereas right-side-out ghosts did not affect the clotting time. Fig. 1 is a representation of a biological membrane and shows the basic structural unit made up of a lipid bilayer containing proteins and glycoproteins. Fig. 2 shows the specific distributional asymmetry of phospholipids within the membrane of platelets and red blood cells.

The characteristic feature of biological membranes is their functional and compositional lipid asymmetry which has been described in several cell types and is thought to stem from the requirement of biological membranes to have asymmetric protein distributions across the bilayer [8]. In all of the cells for which lipid compositional asymmetry has been described, negatively charged phospholipids such as phosphatidylserine are found predominantly on the inner, cytoplasmic side of the membrane, whilst the neutral, zwitterionic PC-containing lipids predominate in the outer leaflet. The negatively charged phospholipids, such as phosphatidylserine, are thrombogenic and it is thought that this membrane asymmetry may serve a biological purpose in the maintenance of the delicate balance between haemostatis and thrombosis [9–12].

Fig. 2. **Schematic representation of the asymmetry in the distribution of the major phospholipids within the plasma membranes of human platelets and erythrocytes**

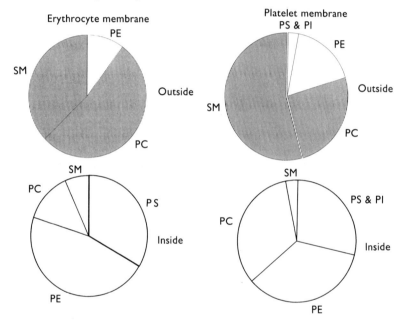

Shown are the percentages of the total phospholipid in each monolayer. Abbreviations used: PC, phosphatidylcholine; SM, sphingomyelin; PE, phosphatidylethanolamine; PS, phosphatidylserine; PI, phosphatidylinositol. Adapted from [4] with permission.

In *in vitro* tests, in which various phospholipid coatings were applied to surfaces, Dennis Chapman's group showed that the negatively charged phospholipids exhibited a very high procoagulant activity [4]. These coatings were shown to activate the intrinsic coagulation pathway, leading to clot formation. This was in stark contrast to the PC-containing surfaces which were not active in coagulation tests and were therefore non-thrombogenic, as would be expected for a major component of the outer surface of an erythrocyte [4, 13, 14].

Dennis Chapman proposed that by reproducing the interfacial characteristics of the outer surface of the lipid matrix of red blood cell and platelet biomembranes, it would be possible to produce surfaces that are non-thrombogenic and therefore biocompatible [4]. He proposed that the mimicry of the non-thrombogenic outer-cell surface of the biological cell membrane by using the PC group for coating, or derivatizing existing plastics and polymers, would lend itself to applications in the biomaterials field. This would lead to the production of a new generation of biomedical devices with greatly reduced potential for thrombogenesis. In order to further develop the concept of biocompatibility through biomembrane mimicry, Dennis Chapman founded a company, Biocompatibles Ltd, which was launched in 1987. This company has focused on the development of PC coatings and PC-containing materials for use in the manufacture of implantable and non-implantable medical devices. The first commercial products containing PC surface treatments for enhancement of biocompatibility are expected to be launched before the middle of this decade.

When present in a close-packed, ordered, planar array such as in the form of a phospholipid coating at an interface, the PC headgroup exhibits very low levels of interaction with proteins and cells. Evidence for this can be drawn from the many studies that have been carried out using phospholipid liposomes as a model system to study phospholipid interactions with biological solutions [5, 6]. Some analogies may be made between these studies and what might occur when phosphatidylcholine or other molecules possessing the PC headgroup are presented in the form of an organized layer at an interface. A surface containing phosphatidylcholine in a smooth, planar and pure, tightly packed form with the phospholipids in the gel phase will not be penetrated readily by proteins. This inference was made independently by Scherphof et al. [5] and Bonté and Juliano [6]. It is apparent therefore from these and other studies that, in order to minimize or eliminate plasma protein interactions with a PC-containing surface, fairly close apposition of the headgroups is required, together with the absence of molecules that can cause irregularities or packing defects in the array. Gel-phase phosphatidylcholine bilayers with a low degree of curvature have the lowest levels of interaction.

Some investigations on the interaction of liposome model systems with the cellular elements of blood have been reported [15, 16]. It was shown that phosphatidylcholine liposomes do not induce the aggregation of platelets, in either plasma or blood. On the other hand, positively and negatively charged liposomes have been shown to interact with platelets [15, 16].

Recently, Nakabayashi and coworkers have synthesized polymers from a PC-methacrylate derivative copolymerized with either methyl or butyl methacrylate [17]. Using *in vitro* haemocompatibility assays they have demonstrated that membranes incorporating this material are non-thrombogenic [18].

Table 1. The contact angles measured using a sessile drop method on two typical substrates used in medical devices

Substrate	Contact angle		
	Control	DAPC	DPPC
Polyethylene ribbon	80°	10°	15°
PVC hard tubing	79°	14°	<5°

The coatings used were of DPPC and DAPC.

Phosphorylcholine surface coatings

The PC headgroup is zwitterionic and has a very high binding affinity with water. Materials coated with a PC derivative are rendered extremely hydrophilic and one hypothesis for why PC is biocompatible is that when a protein approaches a close-packed PC-containing interface, it is confronted by an inpenetrable sea of bound water which prevents interaction with the surface. This can be illustrated by the calculation of contact angles of materials prior to and following coating with PC derivatives. Table 1 shows the difference in contact angle between uncoated and PC-coated materials. The contact angle of the surface of polyethylene and polyvinylchloride is dramatically reduced following coating by either dipalmitoyl-phosphatidylcholine (DPPC) or a polymerized diacetylenic phosphatidylcholine (DAPC) derivative. Moreover, X-ray diffraction studies have shown that immobilized phospholipid layers on surfaces have a lamellar configuration and are therefore present as tightly packed, membrane-like, stable arrays [19].

Polymeric DAPC derivatives with long acyl chains, high phase-transition temperatures and tight molecular packing have been used to produce stable, haemocompatible coatings [4, 13, 14]. These derivatives can be coated onto a material from solution in the monomeric form and subsequently polymerized to form a stable, cross-linked matrix at the material interface. Polymerization can be carried out using either the short wavelength component of ultraviolet light, or by γ irradiation. This is illustrated in Fig. 3. The latter method has the added advantage that sterilization of a medical device coated in this way is carried out simultaneously.

Spectroscopic studies have shown that, upon polymerization, the two conjugated triple bonds of the monomer are replaced by an alternating double- and triple-bonded, conjugated structure [20, 21]. Various DAPC derivatives have been synthesized. In order to tailor the properties of a surface film of this material to a particular substrate, the acyl chain length can be varied or one of the diacetylenic acyl chains can be replaced with a

Fig. 3. Formation of the polyconjugated phospholipid polymer from the diacetylenic monomer

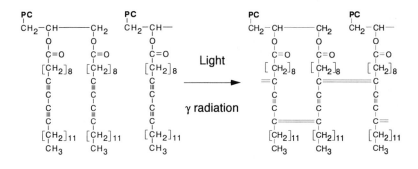

saturated fatty acid. Polymerization efficiency is greatest when the molecular lattice in its monomeric form is ordered and tightly packed [22].

The successful application of PC-based coatings to the surface of a medical device is reliant upon favourable interfacial interactions, whether the coating is chemically attached or physically adsorbed. The performance of the coating *in vivo* is also reliant upon favourable interactions, but in this case between the coating and a biological fluid. Hence, detailed analyses of PC surface coatings using surface analytical techniques is important and necessary. The main techniques employed have been X-ray photoelectron spectroscopy (XPS), time-of-flight secondary ion mass spectroscopy (ToFSIMS), scanning electron microscopy (SEM) and contact angle measurements [23–25]. Of these techniques SEM is used predominantly for the study of the interaction of a biological fluid with a surface, for example to observe platelet interaction with a surface following blood contact. The other techniques mentioned above have also been employed to study such interactions [23], but to a much lesser extent than SEM.

Although SEM can be used for the study of the morphology of surface coatings, in general more useful information can be obtained using XPS and ToFSIMS. In XPS, a focused beam of AlKα X-rays bombard a sample. This initiates the generation and emission of photoelectrons from the sample surface. These photoelectrons have differing kinetic energies which are characteristic of both the orbital and the parent atom. It is therefore possible by using this technique to determine all the elements that are present at a material interface, except for hydrogen and helium. In ToFSIMS a polymer surface is bombarded by accelerated ions which, by a process of energy transference to the surface atoms or molecules, can induce bond breakage at the material interface. By detecting the intensity and mass of the surface-ejected (sputtered) ions, the nature of the surface species can be inferred. This technique stands out among the surface spectroscopies for its high sensitivity, which is in the parts per billion range in ideal cases.

Figs. 4–6 and Tables 2 and 3 show the type of data that can be obtained using the XPS and ToFSIMS techniques when applied to DPPC and DAPC coatings [24, 25]. It has been found that a general pattern emerges for phospholipid coatings. Fig. 4 is a typical wide-scan XPS

Fig. 4.   A wide-scan XPS spectrum of a typical DPPC-coated surface

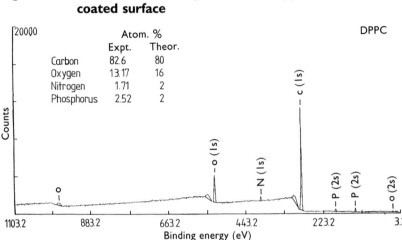

(Courtesy of M. C. Wiles.)

spectrum for a DPPC coating (silver was used as substrate in this case to obtain a model surface) and shows an elemental composition consistent with the presence of the PC headgroup. The information obtained pertains to the outer 4 mm of the surface. The ratios of the elements measured are close to the predicted theoretical values, thus indicating that the surface is free from either hydrocarbon or other contamination.

The slightly higher values for carbon and phosphorus compared with that of nitrogen can be put down to effects associated with the orientation of the phospholipid at the surface, which had not been hydrated prior to analysis. In this situation the PC headgroup is adjacent to the substrate, with the acyl chains pointing outwards towards the XPS detector. This situation would obviously change upon hydration of the coating, as the contact angles obtained for PC-coated surfaces (Table 1) suggest the coating is hydrophilic when hydrated (as opposed to hydrophobic, as would be expected when the coating is in the dry state).

In addition to elemental composition at a given point, XPS analysis at several points on a sample surface can provide information on coating homogeneity. By varying the angle of incidence of the focused X-ray beam, depth information can also be obtained. The XPS spectrum shown (Fig. 4) is

Fig. 5. A ToFSIMS mass spectrum of the region identifying the fragments caused by the phosphorylcholine headgroup: the example shown is of DPPC

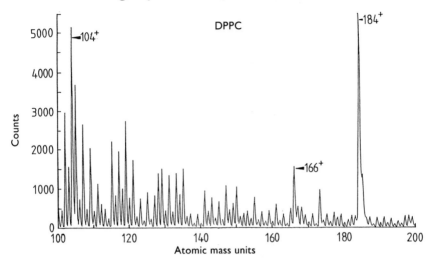

(Courtesy of M. C. Wiles.)

Fig. 6. A ToFSIMS mass spectrum showing the molecular ion of DPPC

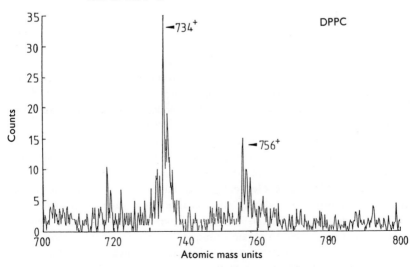

$734 = M + H^+$ and $756^+ = M + Na^+$. *(Courtesy of M. C. Wiles.)*

Table 2. Typical SIMS fragments obtained from the analysis of phosphorylcholine-based treatments

Atomic mass unit	Fragment
79^-	PO_3^-
86^+	$CH_2=CH\overset{+}{N}(CH_3)_3$
104^+	$HOCH_2CH_2\overset{+}{N}(CH_3)_3$
166^+	$\underset{O}{\overset{O}{\diagdown}}P-OCH_2CH_2\overset{+}{N}(CH_3)_3$
184^+	$HO-P-OCH_2CH_2\overset{+}{N}(CH_3)_3$
224^+	$CH_2-O-P-OCH_2CH_2\overset{+}{N}(CH_3)_3$ $\quad\vert\qquad\;\;/\;\backslash$ $\;\,CH\quad HO\;\;O$ $\quad\vert$ $\;\,CH$

(The fragment at 224^+ a.m.u. is only present in glycerophosphatidylcholine coatings.)

typical of both monomeric and polymeric glycerophosphatidylcholine surfaces and it is the phosphorus-to-nitrogen ratio which is the critical parameter for confirming the presence of PC at the interface.

Complementary to XPS is the information that can be obtained using ToFSIMS. Table 2 and Fig. 5 show the information that can be obtained using this technique for any PC surface treatment. Fragments which have been assigned between 79 + to 184 + atomic mass units (a.m.u.) are all characteristic fragments of the PC headgroup. The fragment assigned for 224 + a.m.u. is characteristic of all glycerophosphatidylcholines and is an identifying marker for PC treatments, such as DPPC and DAPC.

For acyl chain-containing glycerophospholipids, additional information can be obtained by measuring the higher-molecular-weight fragments (Table 3) and it may even be possible to measure the molecular ion, as is shown for DPPC in Fig. 6. Hence, by using ToFSIMS it is possible to obtain molecular information about the surface and by monitoring the fragmentation pattern (that is to say the ions formed and their relative intensities), it is possible to obtain a unique 'fingerprint' of the surface treatments being used. The latter situation is important when characteristic fragments cannot be identified for a treatment. In this case it may be possible to identify a treatment simply from the intensity ratios of the PC headgroup fragments. In addition, this type of analysis provides information on whether

Table 3. Characteristic SIMS fragments obtained from ToFSIMS analysis of DPPC and DAPC-coated surfaces

Atomic mass unit	DPPC	Atomic mass unit	DAPC
239$^+$	$^+O{\equiv}C{-}(CH_2)_{14}CH_3$	357$^+$	$CH_3(CH_2)_{11}(C{\equiv}C)_2(CH_2)_8C{\equiv}O^+$
257$^+$	$H_2OOC{-}(CH_2)_{14}CH_3$	790$^+$	$CH_2OCO(CH_2)_8(C{\equiv}C)_2(CH_2)_{11}CH_3$
313$^+$	$CH_2OCO(CH_2)_{14}CH_3$		$\dot{C}HOCO(CH_2)_8(C{\equiv}C)_2(CH_2)_{11}CH_3$
	$^+\dot{C}H$		$\dot{C}H_2^+$
	$\dot{C}H_2OH$	803-6$^+$	$CH_2OCO(CH_2)_8(C{\equiv}C)_2(CH_2)_{11}CH_3$
551$^+$	$CH_2OCO(CH_2)_{14}CH_3$		$\dot{C}HOCO(CH_2)_8(C{\equiv}C)_2(CH_2)_{11}CH_3$
	$\dot{C}HOC(CH_2)_{14}CH_3$		$\dot{C}H_2OH\ (H^+\pm)$
	$^+\dot{C}H_2$	373$^-$	$CH_3(CH_2)_{11}(C{\equiv}C)_2(CH_2)_8COO^-$
255$^-$	$CH_3(CH_2)_{14}COO^-$		
647$^-$	$CH_2OCO(CH_2)_{14}CH_3$		
	$\dot{C}HOCO(CH_2)_{14}CH_3$		
	$\dot{C}H_2OPOH$ $\underset{O\ \ O^-}{/\!/\,\backslash}$		
673$^-$	$CH_2OCO(CH_2)_{14}CH_3$		
	$\dot{C}HOCO(CH_2)_{14}CH_3$		
	$\dot{C}H_2OPOCH{=}CH_2$ $\underset{O\ \ O^-}{/\!/\,\backslash}$		

the coating is intact or has undergone any decomposition during the coating process. In conclusion, from surface analysis it is possible to determine whether a coating process has been successful and whether impurities which might affect coating performance in terms of its biocompatibility are present.

The diacylglycerophosphatidylcholine structure is not a prerequisite for good biocompatibility. The PC headgroup itself, however, is necessary. This was demonstrated by Dennis Chapman and coworkers, who showed that coatings applied by covalent attachment of derivatives of PC to the surface of glass and various polymers were biocompatible [26–28]. XPS and *in vitro* haemocompatibility tests were used to characterize the surfaces produced. Fig. 7 shows the structure of an ethanolamine and a dimethylsilyl chloride derivative of PC which were synthesized and used in the experiments described above. In order to link these derivatives covalently to a substrate, the appropriate chemical groups are required at the surface. Where

Fig. 7. Structures of phosphorylcholine ethanolamine (above) and phosphorylcholine ethylene glycol dimethylsilyl chloride (below)

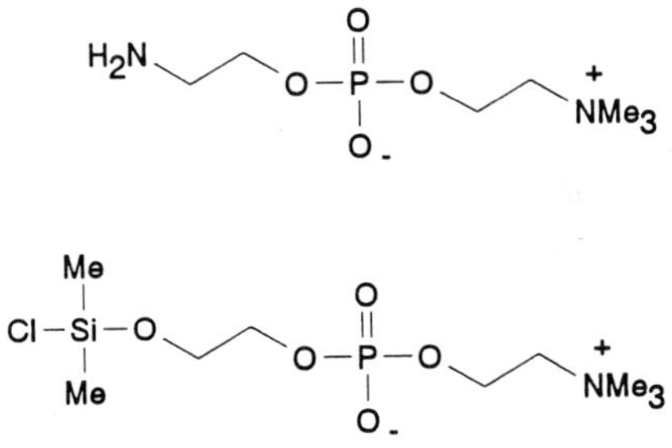

the appropriate chemical functionalities were absent, the substrate material was treated either chemically or by plasma glow discharge to modify the surface in order that reaction to covalently attach the PC derivative could be carried out.

In addition to the above relatively small molecular weight functionally active PC derivatives, polymeric methacrylate-based polymers containing the PC headgroup have been synthesized and attached to a variety of substrates [29–31].

The structure of these PC-containing polymers can be tailored to allow application of a stable coating on a particular substrate. For example, a substrate containing surface carboxylic acid groups allows covalent attachment of a PC-containing methacrylate polymer bearing amino groups. Where suitable chemical groups are not available on a material surface they can be introduced by several possible surface modification methods. This can be achieved chemically, by plasma glow discharge, or by application of a subbing, surface layer containing suitable functionalities. Such methods are commonly used as a prelude to chemical coupling of specific moieties to surfaces [32].

In the case of physiadsorbable polymeric coatings, long acyl chain-containing monomers such as lauroyl methacrylate have been copolymerized with a PC-methacrylate derivative (2-methacryloyloxyethyl phosphorylcholine) to produce polymeric PC derivatives which have been coated onto hydrophobic materials to produce extremely stable surface films with excellent biocompatibility [30, 31, 33].

It is therefore possible to adjust factors which determine the film forming properties and degree of adhesion of the PC polymer to suit the specific interfacial properties of a substrate. The degree of interfacial hydrophobicity or surface free energy of a material can therefore determine the PC polymer properties and composition required to produce a stable coating [29–31].

A further approach to chemical attachment of PC to a surface involves the use of graft polymerization techniques, where polymerization is initiated from the surface of a substrate. Graft polymerization methods involve the formation of free radicals on the substrate surface, which will subsequently initiate polymerization. Many methods are available for achieving this, including radiation- and plasma-induced grafting as well as several chemical grafting methods.

The most efficient method of inducing graft polymerization chemically is to use a redox couple, in which one component is chemically attached to the polymer. For example, grafting of PC onto polymer surfaces containing hydroxyalkyl groups has been achieved by the use of ceric ammonium nitrate [34, 35]. This process is illustrated in Fig. 8. Because cerium also acts as an inhibitor, only short chains are formed (typically 20–250 monomer units) [34]. Using this method, PC-grafting to several substrates has been carried out, including most of the polymers of interest within the biomaterials area, as well as metals such as stainless steel [35, 36]. The success of the cerium-induced grafting process relies on the presence of a suitable functionality at the surface — in this case hydroxyalkyl groups. Where suitable functional groups are absent, they have been introduced using standard surface modification methods, as discussed above (see also [32]). Using a combination of surface modification and chemical grafting techniques, it has been possible to attach PC-methacrylate to a variety of substrates [34, 35]. Subsequent haemocompatibility tests on polymers treated in this way show

Fig. 8. Schematic diagram showing the cerium-induced route for grafting of PC-methacrylate to hydroxyl-containing polymers

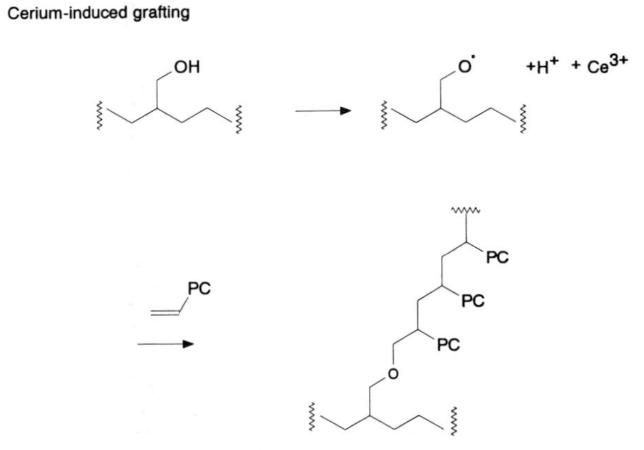

Cerium-induced grafting

$+H^+ + Ce^{3+}$

(Courtesy of T. O. Glasbey.)

a marked reduction in thrombogenicity, relative to the uncoated controls [34, 35].

Chemical grafting of PC-methacrylate to polyurethane surfaces has been achieved by the use of isocyanate reagents, notably haloalkyl or haloacyl isocyanates, or isocyanatoethyl methacrylate [34–36]. Fig. 9 is a schematic diagram, showing a route used for grafting of PC to polyurethane. Coupling has been effected both homogeneously and heterogeneously, and *in vitro* biocompatibility tests have been carried out to assess the potential performance of these materials in biomedical applications [34]. As with the cerium-grafted samples, the *in vitro* haemocompatibility of polyurethanes was shown to be greatly improved following graft polymerization of PC-methacrylate.

PC can not only be used to coat the surface of materials, but in suitable forms can be incorporated throughout the bulk of a polymer to produce new materials which are inherently biocompatible [37–39]. A prerequisite for such materials, in addition to good biocompatibility, is appropriate mechanical and physical properties which would allow fabrication of a device. New materials have been produced by the introduction of suitable PC derivatives as plasticizers into polymers such as PVC [37] and by

Fig. 9. Schematic diagram showing possible routes for grafting of PC-methacrylate to polyurethane

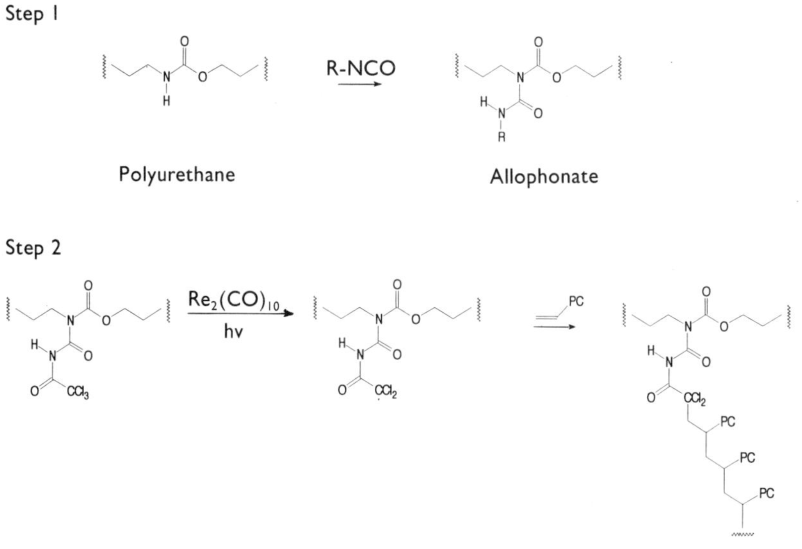

Step 1, activation of urethane linkages with isocyanates; step 2, grafting of PC-containing monomer, e.g. when R = COCCl$_3$. (Courtesy of T. O. Glasbey.)

copolymerizing PC monomers into the polymer backbone of polyurethanes and polyesters [38, 39].

Such new materials could in the future have applications as long-term implantable devices where an inherently biocompatible material would retain its efficacy over extended periods (years) of contact with body fluids or in applications where excessive physical stress and wear might preclude the use of a surface coating.

Haemocompatibility of phosphorylcholine coatings

When blood contacts a thrombogenic surface the haemostatic system, especially the coagulation process, is immediately activated. The surface rapidly acquires a layer of adsorbed plasma proteins, the nature of which mediates subsequent thrombotic events. At least 60% of this adsorbed protein layer is fibrinogen, which is present in blood at a concentration of between 2 and 4 mg/ml. The structural characteristics and limited solubility of fibrinogen are thought to predispose it to binding to foreign surfaces. Following adsorption of fibrinogen, platelet adhesion and activation occurs. This results in the secretion of further fibrinogen from activating platelets, together with factors which in turn effect additional platelet activation. An insoluble network is formed by the conversion of fibrinogen to fibrin by thrombin, which stabilizes the growing mass of platelets to form a thrombus. Thrombin is generated from prothrombin following activation of the intrinsic coagulation pathway, which is initiated by adsorption to the surface and activation of Factor XII.

The combination of fibrinogen binding, platelet adhesion and activation, and activation of the clotting factor cascade (Fig. 10) leads to the

Fig. 10.   Induction of clot formation by biomaterials via the intrinsic pathway

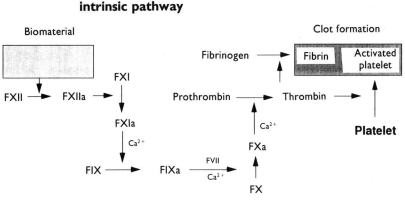

formation of clots at a thrombogenic surface. The thrombogenic process is rapid and localized and does not adversely affect the fluidity of blood elsewhere in the circulation. The events which occur in response to contact of a thrombogenic material with blood are summarized in Table 4.

Many methods are available for the study of the events leading to clot formation, induced by a foreign material in contact with blood. Dennis Chapman and coworkers have carried out clotting studies and thromboelastography measurements to illustrate the biocompatibility of surfaces coated with PC derivatives [13, 14]. This method allows measurement of the elastic properties of whole blood clots as they are formed. It uses a steel piston placed in a cuvette having a 1 mm clearance between the surfaces. Blood is placed within the cuvette. The cuvette oscillates through 4°45' in a 10 s cycle giving a shear rate of 0.28 per second. The piston is suspended by a torsion wire linked to a chart recorder. The torque of the cuvette is transmitted to the piston via the fibrin strands of the blood clot as coagulation proceeds. The resulting trace is called the thromboelastogram (TEG), which records the elasticity of the formed blood clot [40, 41]. A normal TEG trace, together with measurable parameters, is shown in Fig. 11. Each of the parameters described in the legend to Fig. 11 reflects a different component of whole blood coagulation.

Fig. 12(a) shows a typical TEG trace for a system in which the plastic cuvette and steel pin are uncoated. As both surfaces are thrombogenic, fibrin strands form readily within the cuvette and the process of clot from formation can be followed easily, as indicated by the recorded trace. Figs. 12(b) and (c) show TEG traces for DPPC and DAPC, respectively. The PC derivatives used in these experiments were coated onto both the plastic cuvette and stainless steel components of the apparatus so that all blood contacting interfaces were coated. The results show clearly that the two PC coatings render the surfaces haemocompatible and control experiments carried out in the same study, using octacosane-coated components in the apparatus, highlight clearly the importance of the PC headgroup as the effective, non-thrombogenic component of the phospholipid molecule, rather than the hydrocarbon (acyl) chain. In the case of octacosane-coated

Table 4. Events occurring in response to contact of a thrombogenic material with blood

Protein absorption	Platelet interactions	Activation phenomena
Fibrinogen Immunoglobulin Albumin C-reactive protein	Adhesion Activation	Clotting cascade — in particular Factor XII and Complement cascade (C3 to C3a)

Fig. 11. A typical thromboelastogram, with the TEG parameters shown

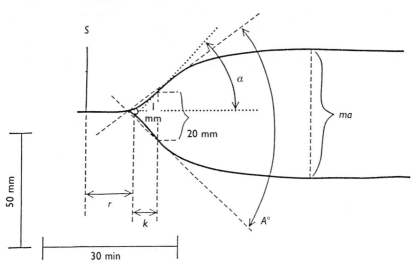

Abbreviations used: S, start mark; r, reaction time; k, clotting speed; ma, maximum amplitude; A°, angle between the first movement of the trace and both shoulders of the trace; α, maximum slope of one shoulder as indicated. Adapted from [13] with permission.

components, the TEG traces produced were very similar to that shown in Fig. 12(a) where all blood contacting surfaces were uncoated.

Fig. 10 and Table 4 illustrate many of the events which occur upon the initiation of thrombogenesis. An examination of the interaction of fibrinogen and platelets with a biomaterial surface *in vitro* can yield valuable information regarding potential haemocompatibility. Recently, sensitive assay methods have been developed for the measurement *in vitro* of fibrinogen adsorption [42], platelet adhesion [33] and platelet activation [43]. These techniques, in combination with SEM and surface analysis have been used extensively within Biocompatibles Ltd to evaluate the performance of PC-coated materials with materials currently used in the biomedical device area [29, 33, 34].

Fibrinogen adsorption measurements can be used as an initial screen of the potential haemocompatibility of a surface since, as was discussed above, this protein binds in large quantities to material surfaces and in so doing affects subsequent events such as platelet adhesion and activation Determination of the amount of fibrinogen adsorbed to materials coated with PC derivatives has been performed using a rapid and sensitive sandwich-type immunoassay method [42]. Small squares of material, 1 cm² in

Fig. 12. Thromboelastogram traces

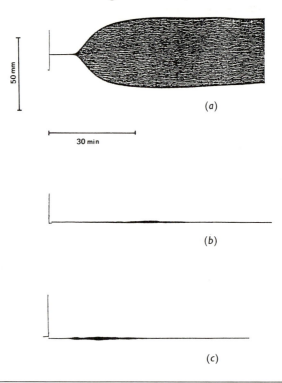

(a) A typical TEG trace for a system where the plastic cuvette and steel pin are uncoated; (b) a TEG trace for DPPC (coated onto both cuvette and pin); and (c) a TEG trace for polymerized DAPC (coated onto both cuvette and pin). Reproduced from [14] with permission.

surface area, are incubated in plasma in wells of a microtitre plate for periods ranging from 10 to 30 min at ambient temperature. After washing, the material is incubated with a specific antibody (goat anti-human fibrinogen) followed by rabbit anti-goat IgG peroxidase conjugate. Following washing and transfer to a second, clean microtitre plate, a chromogen is added and the colour that develops is measured spectrophotometrically. The measured optical density at 650 nm is proportional to the quantity of fibrinogen bound to the surface of the material. The sensitivity of the method allows levels of fibrinogen as low as 0.1 ng to be measured with a high degree of confidence.

Fig. 13 shows the relative differences in fibrinogen adsorption between uncoated polyethylene and polyethylene coated with DPPC, DAPC and lysolecithin (lyso-PC, the palmitoyl derivative). The data shows that the DAPC coating binds considerably less fibrinogen than either uncoated polyethylene or material coated with the other lipids.

It is interesting to note that significantly higher levels of fibrinogen are bound to surfaces coated with lyso-PC, compared with the other coated samples. Lyso-PC coatings have a much higher degree of fluidity compared with the other lipids shown and can therefore be penetrated more readily by

Fig. 13.   The uptake of fibrinogen onto coated and uncoated polyethylene

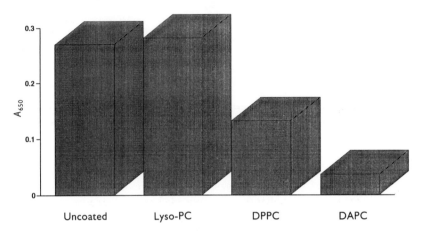

(Courtesy of R. R. C. New.)

proteins. In addition, lyso-PC coatings are unstable (Y. P. Yianni, unpublished work) and are more likely to be lost from the surface, thus exposing the original substrate. The small difference in the level of fibrinogen binding between the DPPC- and DAPC-treated samples is thought to result from the greater stability, lowered surface mobility and tighter molecular packing of the polymerized phospholipid, which binds extremely low levels of fibrinogen.

Table 5 shows the results of fibrinogen adsorption experiments performed on various materials, all coated with DAPC. The results are expressed as percentage reductions in fibrinogen adsorption for the DAPC-coated material relative to the uncoated control. It is evident that a considerable reduction in fibrinogen adsorption can be achieved for a wide variety of medically important materials following coating with a PC derivative. Similar reductions in fibrinogen binding to those observed with DAPC-coated substrates have been found using other derivatives. For example, covalently linked PC derivatives attached by graft polymerization (see above and [34, 35]) give comparable reductions in fibrinogen deposition, as do physically adsorbed PC-polymers produced from copolymerization of methacrylate-containing monomers [30, 31, 33].

Table 5. Results of fibrinogen adsorption experiments performed on various materials coated with DAPC

Substrate	Reduction in fibrinogen adsorption (%)
PVC	90
Polypropylene	83
Polyethylene	86
Polystyrene	90
Polyurethane	69
Polyimide	91
316L Stainless steel	84
Polytetrafluoroethylene	90
Polyamide	84

The results are expressed as percentage reductions in fibrinogen adsorption for the coated material relative to the uncoated control (coefficient of variation is between 5% and 10%).

In addition to surface coatings, PC derivatives can be incorporated throughout the bulk of a polymer and coextruded to produce new materials that are inherently biocompatible. Purified natural lecithins are suitable for this purpose and can be coextruded with, for example, PVC in varying proportions to produce flat sheets or tubing suitable for biocompatibility testing [37]. Fig. 14 shows results of fibrinogen binding experiments carried out on PVC containing varying proportions (% by weight) of egg lecithin (egg-PC). A considerable reduction in the level of fibrinogen binding was

Fig. 14. The uptake of fibrinogen onto PVC polymers containing varying levels of egg-PC

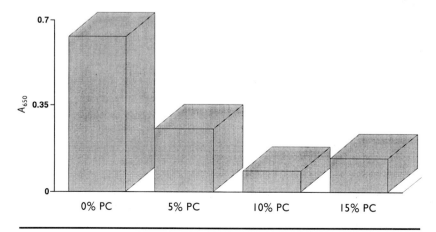

observed even at a level of incorporation as low as 5% egg-PC. Further tests showed that the physical and mechanical properties of the material were retained and that any changes in these properties at higher levels of incorporation of egg-PC could be compensated for by slight modification of the overall polymer formulation. The slight increase in the fibrinogen adsorption levels at 15% egg-PC incorporation are thought to be a result of increased rates of plasticizer migration to the surface of the material.

Study of the interaction of platelets with a biomaterial surface can be a further indicator of the potential haemocompatibility of that material *in vivo* and can be used both to compliment the fibrinogen assay and to provide additional information. Platelets are disc-shaped cell fragments averaging 1.5 μm in diameter and 0.5–1 μm in thickness. They are derived from megakaryocytes and have a primary role in coagulation, providing the surface for molecular assembly of the prothrombinase complex. The study of platelet adhesion and activation to biomaterials is a well accepted method for predicting the potential haemocompatibility of that material *in vivo*. Several methods exist for the study of platelet adhesion, including measurement of the release of platelet factor 4 and β-thromboglobulin which are released following adhesion of a platelet to the surface and its subsequent activation [44]. Three methods have been employed to study platelet adhesion and activation on surfaces coated with PC derivatives. The first is a luminometric assay for determination of platelet adhesion based on the measurement of ATP associated with surface-adhered platelets [33]. The second assay is the detection of platelet activation using an immunoassay that utilizes antibodies directed against a specific receptor present in activated platelets [43]. Finally, SEM has been used to study the morphology of platelets interacting with PC-coated surfaces following contact with blood both *in vitro* and *in vivo*.

Platelet adhesion on PC-coated surfaces has been measured using a luminometric method designed to determine the number of platelets adhered to surfaces by quantifying the ATP associated with them. Samples are initially incubated in fresh citrated human blood at ambient temperature for 30 min. The sample is then removed from the blood and washed carefully to remove loosely adhered material. Platelet-bound ATP is then extracted using a trichloroacetic acid solution and the ATP measured luminometrically using the LKB ATP assay kit. In this assay, ATP is analysed by measuring the light emitted as a result of the enzyme-catalysed conversion of ATP to ADP in the presence of luciferin. The levels of ATP can be related to numbers of platelets bound on a surface using standard curves.

Platelet activation for PC-coated surfaces has been determined using an immunoassay that utilizes antibodies directed against GMP-140 (Fig. 15) which is a membrane protein that is expressed at the surface of a platelet upon degranulation following activation [43]. After binding of the anti-GMP-140 antibody to the activated platelets, a second antibody, sheep anti-mouse IgG peroxidase conjugate, is added, followed by a substrate in order to generate a chromophore. This is measured spectrophotometrically.

Fig. 16 shows the relative differences in platelet adhesion between uncoated polyethylene and polyethylene coated with DPPC, DAPC and lyso-PC. The level of platelet interaction with all coated surfaces is reduced substantially relative to the control. In this experiment, the lyso-PC deriva-

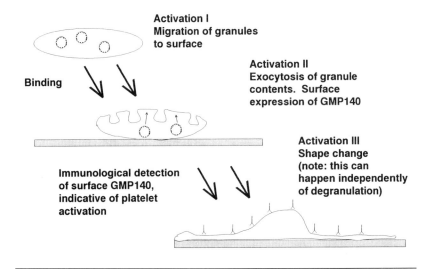

Fig. 15. Principle of the anti-GMP140 assay for the immunodetection of activated platelets on biomaterial surfaces

(Courtesy of R. R. C. New.)

tive exhibited lower levels of platelet adhesion compared with the control, although platelet adhesion for this sample was higher than that observed with DPPC- and DAPC-coated polyethylene. This is in contrast to the results of fibrinogen adsorption tests (Fig. 13), which reveal a higher level of fibrinogen adsorption to the lyso-PC sample than to the control. This is thought to result from headgroup packing defects present in the lyso-PC layer which allow fibrinogen to interact and to therefore penetrate and bind to the surface. Platelets, which are many times larger than a fibrinogen molecule, cannot bind and penetrate, however, since the PC concentration at the surface is sufficiently high to prevent their adherence. In addition, the lyso-PC surface layer is unstable and would be expected to fail this assay if a long incubation time is used, as a result of loss of coating from the surface (incubation time of the materials with blood was 30 min).

Table 6 shows typical reductions in platelet adhesion obtained for various substrates coated with DAPC. The results are expressed as percent reductions relative to the uncoated material. The right hand column in the table indicates that platelet activation was not detected for any of the coated substrates. The results are in line with those obtained for fibrinogen deposition (Table 5) and show very large reductions in platelet adhesion to substrates coated with PC derivatives.

Fig. 16. **The adhesion of platelets to uncoated and coated polyethylene**

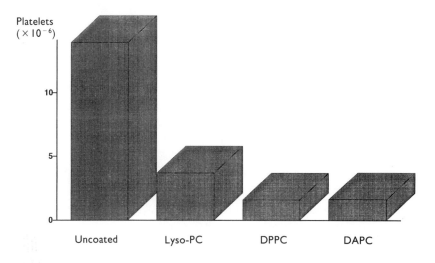

(Courtesy of R. R. C. New).

Fig. 17 shows the results of platelet adhesion experiments carried out on PVC, containing within its structure varying proportions of egg-PC. As with the fibrinogen adsorption results shown for these samples (Fig. 14), the platelet adhesion levels are greatly reduced. This illustrates that, when incorporated at low levels within the structure of a material by coextrusion, PC is present at a sufficiently high surface concentration to substantially improve the haemocompatibility of PVC, which in the conventional form is highly thrombogenic. The above observations demonstrate clearly the improvement in the *in vitro* haemocompatibility of surfaces coated with PC derivatives.

SEM has been used in conjunction with the above tests to confirm these findings and the results from all of these tests correlate very well together. Fig. 18 shows a series of scanning electron micrographs at low and high magnification (scales indicated on the micrographs) showing the surface of uncoated- and DAPC-coated polyethylene following contact with fresh, citrated human blood (incubated with the samples for 30 min at ambient temperature). Fig. 18(*a*) is a low magnification micrograph showing uncoated

Table 6. Results of platelet adhesion experiments performed on various materials coated with DAPC

Substrate	Reduction in platelet adhesion (%)	Platelet activation
PVC	99	No
Polypropylene	86	No
Polyethylene	99	No
Polystyrene	80	No
Polyurethane	97	No
Polyimide	96	No
316L Steel	97	No
Polytetrafluoroethylene	90	No
Polyamide	90	No

The results are expressed as percentage reductions in platelet adhesion for the coated material relative to the uncoated control (coefficient of variation is below 5%). The right hand column in the table indicates whether platelet activation was detected.

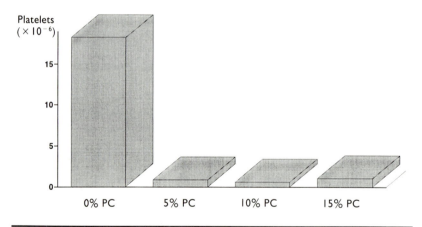

Fig. 17. Platelet adhesion onto PVC polymers containing varying levels of egg-PC

polyethylene. Extensive platelet adhesion can be seen. At higher magnification (Fig. 18b) it is clear that the platelets are activated. Platelets are usually disc-shaped but, when activated, undergo a morphological change which is accompanied by the formation of pseudopodia (see Fig. 18). Figs. 18(c) and (d) show the surface of coated polyethylene following blood contact. The surface is completely devoid of platelets, indicating that the haemocompatibility of the sample has been improved following coating with DAPC.

Figs. 19(a) and (b) are low magnification micrographs of uncoated and DAPC-coated polytetrafluoroethylene respectively. The blood in contact with the sample, although originally anticoagulated with citrate, was recalcified prior to incubation with the samples in order to restore the

Fig. 18. A series of scanning electron micrographs at low and high magnification showing the surface of uncoated and DAPC-coated polyethylene following contact with fresh citrated human blood

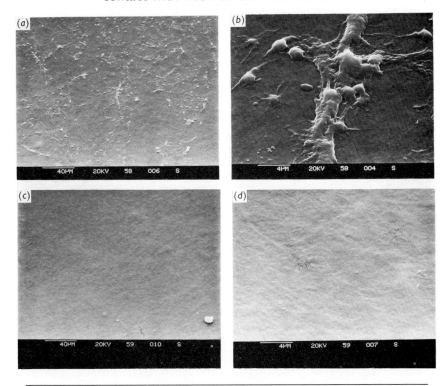

(a) Low magnification micrograph of uncoated polyethylene; (b) high magnification micrograph of uncoated polyethylene; (c) low magnification micrograph of DAPC-coated polyethylene; and (d) high magnification micrograph of DAPC-coated polyethylene. (Courtesy of A. M. Sullivan.)

clotting potential. The uncoated sample (Fig. 19a) shows a thrombus formed on the surface. This consists of an extensive cross-linked fibrin network containing entrapped cells. The PC-coated sample is, in contrast, very clean in that no cellular or proteinaceous deposits can be observed. The surface morphology of the substrate itself is clearly visible.

In addition to electron microscopy, it is possible to observe differences in the haemocompatibility of coated- and PC-coated materials macroscopically following contact with human blood. Fig. 20 shows uncoated and PC-coated PVC tubing following contact with fresh human blood. The

Fig. 19. Scanning electron micrographs of uncoated and (DAPC)-coated polytetrafluoroethylene following contact with citrated blood which was recalcified prior to contact with samples

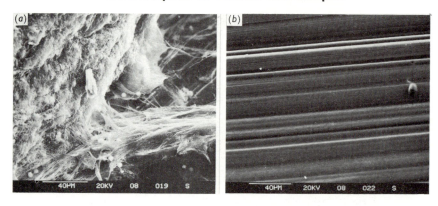

Low magnification micrographs of uncoated (a) and DAPC-coated (b) polytetrafluoroethylene. (Courtesy of E. J. Campbell).

blood had been citrated when drawn and was recalified prior to contact with the surface. Incubation of tubing with blood was carried out at 37 °C for a period of 90 min. The tubing was rotated periodically during incubation. The blood was subsequently drained from the tubes, which were washed with physiological saline prior to being photographed. The lower tube in Fig. 20 is uncoated and coagulated blood is clearly evident. The coated tube (top) contains no clots or other deposits on the surface, showing that the thrombogenicity of PVC tubing can be reduced substantially following coating with PC derivatives.

The results obtained in *in vitro* experiments have been confirmed in an *ex vivo* shunt in a calf model, where coagulation patterns in a left heart bypass system (consisting of PVC tubing and cannulae) coated with DAPC, were compared with a heparinized system [45]. The drug heparin is an anticoagulant and has been used as an anti-thrombogenic coating for medical devices [46].

In this study three calves were used for evaluation of each coating and the left heart bypass for each animal was maintained for 6 h. Coagulation studies included measurement of platelet counts, activated coagulation time, thrombin time, fibrinogen, anti-thrombin III, fibrinopeptide A and SEM of blood-exposed surfaces. No systemic anticoagulant was administered during this study and the results suggested that blood-exposed surfaces coated with a physically adsorbed DAPC coating are at least as thromboresistant as those

Fig. 20. DAPC-coated PVC tubing (above) following contact with fresh human blood (recalcified prior to incubation with the sample), together with an uncoated control specimen (below)

coated with ionically bound heparin. Ionically bound heparin coatings are, however, unstable and gradually lost from the surface when subjected to high shear forces (the blood flow rate used in the above experiments was approximately 3.5 l per min). DAPC coatings have been shown to be stable over such periods under these conditions. Electron microscopy carried out on samples of tubing from the circuit showed that surfaces did not contain deposits of platelets or proteinaceous material [45]. These results correlate very well with the *in vitro* data presented above and suggest that *in vitro* tests can indeed be used as an initial screen to give an indication of the potential haemocompatibility of PC-coated materials in an animal model. The results of the calf model experiments showed that all of the coagulation parameters measured remained virtually unchanged and therefore within the normal range over the course of the experiment.

The ultimate challenge for any technology is in the final application and, in the case of PC-coated biomedical devices, this would involve human clinical tests. Some *in vivo* human work has been carried out using PC-coated PVC thoracic drain catheters (S. Hunter & G. D. Angelini, personal communication). The aim of this study was to investigate the effect of coating with a PC treatment on pericardial blood drainage following open heart surgery. The study was conducted on 97 patients, divided into two groups, one of which received conventional catheters, the other PC-coated catheters. Data collected during the 24 h period in which catheters were inserted showed clearly that the coated drains performed better than the conventional ones. Improvements were seen in the haemocompatibility of the PC-coated devices resulting in significantly reduced levels of clot left in the drains at the time of removal and fewer blockages. In addition, drainage of blood from the chest cavity was more efficient for the PC-coated group, resulting in a lowered incidence of complications such as arrhythmia.

It was stated above that protein interactions at foreign interfaces are the initial events in the biological response to a foreign material. The large reduction in levels of such biointeractions with PC-coated materials can be exploited in areas other than blood contacting applications. Medical devices which come into contact with tissues and biological fluids other than blood may, in the future, benefit from coating with PC derivatives, as may drug delivery systems, clinical assay kits and biosensors. PC-coated materials may also be of value as non-fouling membranes in industrial bioprocesses such as downstream separation and purification systems.

An additional area of application for PC treatments currently being developed at Biocompatibles Ltd involves the use of PC coatings in hydrogel contact lenses [47]. Current lens materials suffer from problems of formation of deposits from tear fluids within the hydrogel structure. This can lead to significant problems for wearers of these devices, who may suffer considerable discomfort, which in some cases may preclude their use. A significant reduction in the level of protein adsorption has been observed in studies of tear protein interactions with hydrogel contact lens polymers with which PC derivatives have been covalently reacted [47]. The proteins studied were lysozyme and albumin. In addition, it has become evident that the hydrophilic nature of the PC headgroup makes it an ideal component of hydrogel systems.

In the coming years it is expected that several products coated with PC will become widely available. Dennis Chapman's original hypothesis has now been developed to the point where clinical studies are showing the significant benefit that biomimetic coatings can confer upon biomedical devices to improve patient care and quality of life.

PC coatings have been shown to be stable, in that the material does not leach from surfaces. In addition, toxicological testing of materials coated with the PC derivatives described above has been carried out. Results of cytotoxicity, acute systemic toxicity, sensitization and irritation tests all indicate that the coatings are non-toxic (Y. P. Yianni, unpublished work commissioned by Biocompatibles Limited and carried out at Huntingdon

Research Centre, U.K.). Finally, PC-coated materials can be sterilized by all current methods used in the medical device industry, without any loss of biocompatibility.

The author gratefully acknowledges Mrs A. V. Hacker of Biocompatibles Ltd for permission to present some of the data. The author would also like to thank the following staff of Biocompatibles Ltd for their valuable contributions: Mr E. J. Campbell, Dr S. A. Charles, Dr T. O. Glasbey, Dr B. Hall, Dr R. R. C. New, Ms A. M. Sullivan and Dr M. C. Wiles. In addition, the author thanks Professor Dennis Chapman for helpful discussions during preparation of this manuscript.

References

1. Mason, R. G., Kim, S. W., Andrade, J. D. & Hakim, R. M. (1980) Trans. Am. Soc. Artif. Intern. Organs 26, 603–610
2. van Berlo, A. & Ellens, D. J. (1988) in Biotechnological Applications of Lipid Microstructures (Gaber, B. P., Schnur, J. M. & Chapman, D., eds.), pp. 341–358, Plenum Press, New York
3. van Oeveren, W., Kazatchine, M. D., Descamps-Latscha, B., Maillet, F., Fischer, E. & Carpentier, A. (1985) J. Thorac. Cardiovasc. Surg. 89, 888–889
4. Hayward, J. A. & Chapman, D. (1984) Biomaterials 5, 135–142
5. Scherphof, G. L., Damen, J. & Wilschut, J. (1984) in Liposome Technology, Volume III (Gregoriadis, G., ed.), CRC Press, Boca Raton
6. Bonté, F. & Juliano, R. L. (1986) Chem. Phys. Lipids 40, 359–372
7. Zwaal, R. F. A., Comfurius, P. & van Deenen, L. L. M. (1977) Nature (London) 268, 358–360
8. Op den Kamp, J. A. F. (1979) Annu. Rev. Biochem. 48, 47–71
9. Zwaal, R. F. A. & Bevers, E. M. (1983) in Subcellular Biochemistry, Vol. 9 (Roodyn, D. B., ed.), pp. 299–327, Plenum Press, New York
10. Zwaal, R. F. A. (1978) Biochim. Biophys. Acta 515, 163–205
11. Hemker, H. C., Lindhout, M. J. & Vermeer, C. (1977) Ann. N.Y. Acad. Sci. 283, 104–110
12. Zwaal, R. F. A. & Hemker, H. C. (1982) Haemostasis 11, 12–39
13. Bird, R. le R., Hall, B., Chapman, D. & Hobbs, K. E. F. (1988) Thromb. Res. 51, 471–483
14. Hall, B., Bird, R. le R., Kojima, M. & Chapman, D. (1989) Biomaterials 10, 219–224
15. Juliano, R. L., Hsu, M. U., Peterson, D., Regen, S. L. & Singh, A. (1983) Exp. Cell Res. 146, 422–427
16. Bonté, F., Hsu, M. J., Papp, A., Wu, K., Regent, S. L. & Juliano, R. L. (1987) Biochim. Biophys. Acta 900, 1–9
17. Ishihara, K., Ueda, T. & Nakabayashi, N. (1990) Polymer J. 22, 355–360
18. Ishihara, K., Aragaki, R., Ueda, T., Watenabe, A. & Nakabayashi, N. (1990) J. Biomed. Mater. Res. 24, 1069–1077
19. Nagumo, M., Ratna, B. R., Qadri, S., Naciri, J. & Shashidhar, R. (1991) Mol. Cryst. Liq. Cryst. 206, 123–130
20. Pons, M., Johnston, D. S. & Chapman, D. (1982) J. Polym. Sci., Polym. Lett. Ed. 20, 513–520
21. Pons, M., Villaverde, C. & Chapman, D. (1983) Biochim. Biophys. Acta 730, 306–312
22. Baughman, R. H. & Yee, K. C. (1978) Macromol. Rev. 13, 219
23. Glasbey, T. O., Paul, A., West, R. & Wiles, M. C. (1992) Int. Symp. Surface Properties Biomaterials, May 11–13, UMIST, Manchester
24. Davies, M. C. & Lynn, R. A. P. (1990) Crit. Rev. Biocomp. 5, 297–341
25. Paul, A., West, R., Wiles, M. C. & Wotherspoon, G. (1992) 4th World Biomaterials Congr., April 24–28, Berlin
26. Durrani, A. A., Hayward, J. A. & Chapman, D. (1986) Biomaterials 7, 121–125
27. Hayward, J. A., Durrani, A. A., Shelton, C. J., Lee, D. C. & Chapman, D. (1986) Biomaterials 7, 126–131
28. Hayward, J. A., Durrani, A. A., Lu, Y., Clayton, C. R. & Chapman, D. (1986) Biomaterials 7, 252–258

29. Chapman, D. & Charles, S. A. (1992) Chemistry in Britain 28(3), 253–256
30. Biocompatibles Ltd, Br. Patent (1991) GB 9114619.1
31. Biocompatibles Ltd, Br. Patent (1991) GB 9117170.2
32. Fettes, E. M. (1964) Chemical Reactions of Polymers (Fettes, E. M., ed.), Wiley Interscience, London
33. New, R. R. C. (1992) in Biotechnological Applications of Phospholipids (Cevc, G., ed.), Marcell Dekker, New York, in the press.
34. Al-Lamee, K. G., Campbell, E. J., Wiles, M. C., Glasbey, T. O., Bamford, C. H. & Yianni, Y. P. (1991) 9th Eur. Conf. Biomaterials, 9–11 September, Chester
35. Biocompatibles Ltd, Br. Patent (1991) GB 9118597.5
36. Bamford, C. H., Al-Lamee, K. G., Middleton, I. P., Paprotny, J. & Carr, R. (1990) Bull. Soc. Chim. Belg. 99, 919–929
37. Valencia, G. P. (1985) Eur. Patent 247 114
38. Chapman, D. & Valencia, G. P. (1984) Eur. Patent 199 790
39. Durrani, A. A. (1986) Eur. Patent 275 293
40. Affeld, K., Berger, J., Muller, R. & Becherl, E. S. (1974) Proc. Eur. Soc. Artif. Organs 1, 26–29
41. Carr, S. H., Zuckerman, L., Carrini, J. A. & Vagher, J. P. (1976) Res. Commun. Chem. Pathol. Pharmacol. 13, 507–519
42. Lindon, J. N., McManama, G., Kushner, L., Merrill, E. W. & Salzman, E. W. (1985) Blood 68, 355
43. Campbell, E. J., Wiles, M. C., New, R. R. C. & Charles, S. A. (1992) J. Mater. Res. Soc. Symp. Proc. 252, 229–236
44. Gaffney, P. J. & Philo, R. D. (1981) in Perspectives in Hemostasis (Fareed, J., Mesmore, H. L., Fenton, J. W. & Brinkhous, K. M., eds.), pp. 405–417, Pergamon Press, New York
45. von Segesser, L. K., Olan, M., Leskosek, B. & Turina, M. I. (1992) ASAIO Abstract, 38th Annu. Meet., p. 25, May 7–9, Nashville, Tennessee
46. von Segesser, L. K. & Turina, M. I. (1989) J. Thorac. Cardiovasc. Surg. 98, 386–396
47. Bowers, R. W. J., Driver, M. J., Jones, S. A., Murray, A. C. & Stratford, P. W. (1992) Abstr., Fourth World Biomaterials Cong., April 24–28, Berlin

Subject index

Acholeplasma laidlawii, 61–63
Actin, 144
Activation energy, 116
Acyl chain, 69
2-O-Acyl-1-O-polyprehnyl-α-D-glucopyranoside, 59–61
Aggregated band 3, 146
Aggregation number, 71
Alamethicin, 174
Alkene, 31
Alpha structure, 169
Amide, 167
   Secondary structure analysis, 166, 167
   i.r. spectrum, 161
Amphipathic helix, 155
Anhydrous soap system, 2
Anionic membrane lipid, 66–69
Anionic phospholipid
   electrostatic repulsion, 95
   phase transition, 69
Anisotropic motion of spin-probe, 86–90
Ankyrin, 141, 144–147
Annulus lipid, 90
Antipeptide antibody, 14–16
Atriopeptin III, 157

Bacillus megaterium, 63, 64
Bacteriorhodopsin
   photocycle, 174
   protein reconstitution, 47
   protein rotational diffusion, 3
   structure, 171
   transient dichroism, 140
Band 3 protein
   crosslinking, 148
   cytoplasmic domain, 147
   oligomeric state, 145
   rotational diffusion, 137–150
   rotational mobility, 146
   self-association, 145, 146
BBMV (see Brush border membrane vesicle)
Beta-structure, 169
Bicontinuous cubic phase
   n.m.r. diffusion study, 72, 73
   structure, 52
   translational diffusion, 70
Bilayer membrane, 104–109
Bile salt, 55
Biocompatible surface, 187–216
Biological membrane
   hydrogenation, 31, 36–44
   non-lamellar phase, 51
Biomembrane
   mimicry, 187–216
   structure, 124–133
Biosynthetic pathway, 38, 39
Boundary lipid, 82, 88
Boundary phospholipid, 109
Browning rotational diffusion model, 71
Brush border membrane vesicle
   cholesterol absorption, 7–18
   lipid-exchange protein, 7
   papain treatment, 11
Bulk-phase lipid, 26
^{13}C Fourier transform n.m.r., 128
^{13}C magic angle sample spinning n.m.r., 127
^{13}C n.m.r., 127
Ca^{2+}-ATPase, 25, 79, 89, 177–179

Caged ATP, 178
Calcium pump activity, 37
Catalyst
  accessibility, 35
  hydrogenation, 31
Cation pump, 36, 37
Cell-surface antigen, 44
Cerebroside orientation, 124, 126
CH_2 asymmetric stretching vibration, 94
Chemical coupling, 198
Chemical modification
  homogeneous catalyst, 29–36
  lipid phase behaviour, 29–48
Chemical structure of lipid phase, 53, 54
Chilling sensitivity, 41
Chlorophyll a fluorescence, 42
Cholate, 55
Cholera toxin
  B subunit, 159
  membrane receptor, 156–159
Cholestane, 11
Cholesterol
  absorption, 7–18
  collision-induced transfer, 16
  desorption, 17, 113, 116
  efflux, 114
  exchange, 109–117
  lateral packing density, 114
  lipid chain dynamics, 3
  membrane fluidity, 95
  molecular packing, 104
  ordering effect, 106
  passive diffusion, 15
  phase equilibria, 55
  phosphatidylcholine exchange, 14
  phospholipid hydrogenation, 33
  –phospholipid interaction, 19–27, 103–117
  rate-limiting desorption, 115
  transbilayer movement, 110
  transfer, 111
  transfer protein, 12
  slow phase, 17
  van der Waals interaction, 107
Cholesterol/phosphatidylcholine bilayer, 107, 108
α-Chymotrypsin
  deconvolved spectrum, 169
  i.r. spectrum, 170
Circular dichroism, 165
Closed micelle, 71
Clostridium butyricum, 63
Clotting factor cascade, 200
Clotting time, 188
Cold stress, 40, 41
Collision-induced lipid transfer, 16
Condensing effect, 124
Contact angle, 196
Counterion association, 66
Cross polarization, 130, 132
Crosslinking of band 3, 148
Cryoprotectant, 54
Cubic lipid phase, 51–74
Cytochalasin B, 181
Cytochrome b_6–f complex, 43
Cytochrome c
  membrane-induced protein destabilization, 159–163
  denaturation temperature, 162
  haem group, 163
  membrane bound, 160
  reversible folding, 162
  solvent accessibility, 161
Cytoskeleton, 141, 143–145

Deconvolved spectrum, 169
Delayed fluorescence, 140
Desorption, 17
Detergent, 55
Deuterium n.m.r. spectroscopy, 3
Diacetylenic acyl chain, 192
Diacetylenic derivative, 45
Difference absorption, 167

Difference i.r. spectrum, 177, 178
Digalactosyldiglyceride, 127
Dioleoyl phosphatidic acid, 68
Diphenylhexatriene, 92, 93
Diphosphatidylglycerol, 68, 69
Dipolar interaction, 119, 120
Double bond, 20–23
Drug vector, 47

Efflux, 114
Elastic free energy of curvature, 68
Electron double layer, 160
Electron
   delocalization, 1
   diffraction, 171
Electrostatic property, 95, 96
Electrostatic repulsion, 67, 95
Encapsulation of drug, 47
Endogenous phospholipase A, 41
Enthalpy of melting process, 83
Eosin-labelled band 3, 144–146
Erythrocyte
   cytoskeleton, 141
   glucose transporter, 181, 182
   membrane structure, 141, 142
   rotational diffusion, 137–150
Escherichia coli, 65
E.s.r. spectroscopy, 1, 86–90

Factor analysis of i.r. spectra, 170
Ferricytochrome c, 161
Fibrinogen
   adsorption measurement, 202
   binding, 204
Flash photolysis, 137
Fluidity, 2
Fluorescence depletion, 140
Fluorescence depolarization
   diphenylhexatriene probe, 92, 93

Fluorescence recovery after photobleaching (FRAP), 70, 149
Foreign interface, 214
Fourier self-deconvolution, 168
Fourier transform i.r.
   bulk-phase lipid, 26
   cubic phase, 73
   lipid–protein contact, 93
   membrane protein, 166
FRAP (see Fluorescence recovery after photobleaching)
Free energy lipid exchange, 113
Freeze-fracture electron microscopy, 78–80

G protein-coupled receptor, 175
Ganglioside G_{M1}, 158
Gastric H^+/K^+-ATPase, 179, 180
Gel to liquid-crystalline phase transition, 22
Globular protein, 168
Glucose transporter, 181, 182
Glyceraldehyde-3-phosphate dehydrogenase, 147
Glycophorin A, 142
Glycoprotein, 86
Graft polymerization technique, 198, 199
Gramicidin, 56

^{1}H-n.m.r. spectroscopy, 119–124
^{2}H-n.m.r. spectroscopy
   applications, 123, 124
   POPC/cholesterol, 125
   reconstituted systems, 90, 91
Haem group, 163
Haemocompatibility
   assay, 191
   phosphorylcholine coating, 200–215
   polyurethane, 201
Haemostasis, 190
Head group, 58
Hexagonal lipid phase, 51–74

High temperature, 41
High water content, 69–72
Homeoviscous adaptation, 37, 38
Homogenous catalyst
  chemical modification, 29–36
  water-insoluble, 31–33
  water-soluble, 33–36
Hydration, 54
Hydroformylation, 30
Hydrogen
  bonding, 59
  donor, 36
  transfer, 36
Hydrogen–deuterium exchange
  cytochrome c, 161
  erythrocyte, 181
  purple membrane, 174
Hydrogenation
  biological membrane, 31, 36–44
  light-dependent, 36
  lipid, 29
  living cell, 43, 44
  multilamellar dispersion, 35
  phospholipid, 32, 33
  photosynthetic membrane, 42
  plasma membrane, 39, 44
  soluble substrate, 34
  thylakoid membrane lipid, 42
  unsaturated fatty acid, 31
  unsaturated membrane lipid, 30
  unsaturated phospholipid, 34
Hydrophobic amino acid, 155
Hydrophobic mismatch, 85
Hydrophobicity analysis, 165, 166

I.r. linear dichroism, 182
I.r. spectroscopy
  anhydrous soap system, 2
  atriopeptin III, 157
  cholera toxin B subunit, 159
  factor analysis, 170
  membrane protein, 165–182
  protein secondary structure, 166–168
Immobilized lipid, 90
Immobilized probe, 87
In-plane vibration, 167
Integral membrane protein
  hydrophobicity analysis, 166
  lateral mobility, 149
Interdigitated packing, 108
Interfacial hydrophobicity, 198
Intermolecular hydrogen bonding, 59
Intrinsic tryptophan fluorescence, 178
Ionic head-group, 54
Isotropic exchange, 167

Kinetic i.r., 175
Kinetics of cholesterol absorption, 8
K_m, 26

L_α phase
  order parameter, 58, 60
  X-ray diffraction, 58
Lateral diffusion, 149, 150
Lateral mobility, 149
Lateral packing density, 144
Lateral phase separation
  cholesterol/phosphatidyl-choline bilayer, 107
  phosphatidylcholine–cholesterol complex, 109
  phospholipid/cholesterol monolayer, 116
Light-dependent hydrogenation, 36
Light-harvesting chlorophyll–protein complex, 40
Lipid
  antipeptide antibody, 14–16
  biosynthetic pathway, 38, 39
  cation pump, 36, 37

chain dynamics, 3
chemical structure, 53, 54
composition, 61-65
crystalline phase, 62
cubic phase, 51-74
deuterium n.m.r.
    spectroscopy, 3
hexagonal phase, 51-74
hydrogenation, 29
mediator, 98
order and dynamics, 92
packing, 59
phase behaviour, 29-48
phase equilibrium, 63, 64
polar head group distribution, 26
polymerization, 45, 46
polyunsaturated ion, 20
protein interactions, 24-27
saturation, 36, 37
structure modification, 19
surrounding protein structure, 27
topology, 39
translational diffusion, 4
unsaturation, 21, 38
Lipid/water phase diagram, 68
Lipid-exchange peptide, 11-14
Lipid-exchange protein
    brush border membrane vesicle, 7
    purification, 17
Lipid-induced peptide folding, 154-156
Lipid-intrinsic protein interaction, 90-92
Lipid–protein contact, 93
Lipid–protein interaction, 93-96
Lipid–sterol interaction, 134
Lipid–water interface, 95, 96
Lipophilin, 182
Liquid crystal, 129
Liquid crystalline phase, 51
Living cell hydrogenation, 43, 44
Lymphocyte hydrogenation, 43
Lyotropic liquid crystalline state, 51

Macroscopic study, 78
Magic-angle sample spinning n.m.r., 121
Magnetic resonance imaging (MRI), 122, 123
MAS (see Magic angle spinning)
Medical device, 196
Melittin, ???
Melting process, ???
Membrane fluidity, 43, 95
Membrane homeoviscous adaptation, 37, 38
Membrane protein
    i.r. spectroscopy, 165-182
    rotational diffusion, 137-150
Membrane receptor for cholera toxin, 156-159
Membrane stability, 41
Membrane structure, 39
Membrane surface potential, 163
Membrane thickness, 95
Membrane-bound ATPase, 179-181
Membrane-induced protein destabilization, 159-163
Membrane-intrinsic protein
    freeze-fracture electron microscopy, 78-80
    –lipid interaction, 77, 78
    macroscopic study, 78
    –phospholipid interaction, 77-98
Metal complex, 34
Metastable gel phase, 61
Methyl-branched acyl chain, 69
Microbial glucolipid, 56-59
Microbial membrane, 61-65
Mixed-acid lipid, 25
Molecular order parameter, 122
Molecular ordering of head group, 58
Molecular packing, 104

Mono-olein, 72–74
MRI (see Magnetic resonance imaging)
Multilamellar dispersion, 35
Myelin basic protein, 182
Myelin spectrum, 132

N.m.r. diffusion study, 72, 73
N.m.r. spectroscopy
  lipids and membranes, 119–134
  lipid–intrinsic protein interaction, 90–92
  low-molecular-weight protein, 165
Na^+/K^+-ATPase, 180
Nematic liquid crystal, 129
Non-lamellar phase
  anionic membrane lipid, 66–69
  biological membrane, 51, 53
  microbial glucolipid, 56–59

Oleoyl-CoA desaturase, 37
Omeprazole, 179, 180
Opioid peptide, 156
Order parameter
  cubic phase, 73
  ^{13}H-n.m.r., 127
  L_α phase, 58, 60
Ordering effect of cholesterol, 106

Packing
  energy, 68
  lipid molecule, 59
  parameter, 67
Papain, 11
Passive diffusion, 15
Peptide folding, 154–156
Peptide hormone, 154–156
pH, 54
Phase diagram
  cholesterol/phosphatidylcholine bilayer, 108
  protein-rich cluster, 84
Phase equilibrium
  anionic membrane lipid, 66–68
  lipid mixture, 64
  membrane lipid, 63
Phase separation
  intrinsic protein, 78
  protein-rich patch, 79
Phase transition
  gel to liquid-crystalline, 22
  intrinsic protein, 80–86
  phospholipid, 2
  reconstituted system, 80
Phosphatidylcholine
  cholesterol exchange, 14
  diacetylenic derivative, 45
  polyunsaturated chain, 23
  translational diffusion, 73
Phosphatidylglycerol, 39
Phosphine, 34
Phospholipid
  asymmetrical distribution, 190
  –cholesterol interactions, 23, 24
  –cholesterol monolayer, 116
  double bond, 20–23
  dynamics, 2
  hydrogenation, 32, 33
  immobilized layer, 192
  –membrane-intrinsic protein interaction, 77–98
  molecular species, 20
  phase-transition, 2
  physical properties, 20–23
  polyconjugated polymer, 193
  surface orientation, 194
  unsaturation, 19–27
Phospholipid:protein ratio, 65
Phosphorylcholine
  biocompatibility, 188–191
  haemocompatibility, 200–215
  surface coating, 192–200, 213
Photocycle, 174
Photosynthetic electron transport, 41–43

Photosynthetic membrane
  hydrogenation, 42
  structure, 176
Photosystem I, 43
Photosystem II
  inhibition, 43
  reaction centre, 176
  secondary structure, 176
Plasma membrane
  hydrogenation, 39, 44
Platelet
  activation, 209
  adhesion, 207
  aggregation, 191
  −biomaterial surface
    interaction, 206
  −coated surface interaction,
    208
  −surface interaction, 193
Polar head group
  distribution, 26
  hydration, 54
Polarization of phosphorescence,
  140
Polarized attenuated total
  reflection i.r.
  spectroscopy, 181
Polarized i.r., 173, 174
Polymerizable sulpholipid, 46
Polymerization
  lipid, 45−47
  liposome as drug vector, 47
Polypeptide aggregation, 80
Polyunsaturation
  phosphatidylcholine, 23
  optimal activity, 20
Polyurethane, 201
POPC/cholesterol, 125
Porin, 182
Pressure, 54
Procoagulant activity, 190
Protein activity, 98
Protein aggregation
  band 3, 148
  intrinsic protein, 78

Protein concentration, 98
Protein destabilization, 169−163
Protein folding, 153
Protein/lipid patch, 82
Protein mobility, 150
Protein reconstitution, 46, 47
Protein rotation, 138−140
Protein rotational diffusion
  bacteriorhodopsin, 3
  1,2-dipalmitoyl-sn-
    phosphocholine, 85
Protein structure
  i.r. spectroscopy, 166−168
  K_m, 26
  lipid structure, 27
  modulation by lipid
    environment, 153−163
Protein−lipid interaction, 24−27,
  124
Protein−protein contact, 82, 88,
  92, 96, 97
Protein−protein interaction in
  erythrocyte membrane,
  138
Protein-rich cluster, 79, 84
Proton-coupled ^{13}C MAS n.m.r.,
  129
Proton-decoupled Bloch decay
  spectrum, 130
Pseudomonas fluorescens, 65
Purification of lipid-exchange
  protein, 17
Purple membrane
  electron diffraction, 171
  polarized i.r., 173, 174

Quadrupole splitting, 123
Quantitative analysis, 169−171

Raman spectroscopy, 93, 165
Reconstituted system
  ^{2}H-n.m.r., 90
  phase-transition, 80
Red blood cell, 117

Resolution enhancement, 168–169
Reversed non-lamellar phase, 53
Reversible folding, 162
Rhodopsin
    lipid–protein interaction, 25
    i.r. spectroscopy, 175, 176
    protein reconstitution, 47
    rotational diffusion, 137
Rotational diffusion
    band 3, 143, 146
    membrane protein, 137–150

Sarcoplasmic reticulum
    Ca^{2+}-ATPase, 25, 177
    mixed-acid lipid, 25
Scanning electron micrograph, 209–212
Second-derivative i.r. spectrum
    gastric H^+/K^+-ATPase, 180
    globular protein, 168
Segmental order parameter, 62
Self-association of band 3, 145, 146
Site-directed mutagenesis, 174
Smectic liquid crystal, 120
Soluble substrate, 34
Solvent accessibility, 161
Spectral distortion, 170
Spectrin network, 141
Sphingolipid, 54
Sphingomyelin, 115
Spin label, 86, 87, 90
Spontaneous curvature, 74
Steady-state fluorescence polarization, 89
Stearic acid, 87
Sterol, 124
Sugar hydroxyl group–water interaction, 122
Surface charge density, 68
Surface pressure curve, 65
Surface treatment, 196

Taurocholate-mixed micelle, 8
Temperature, 54
Terminal olelin, 30
Three-dimensional density map, 171
Thromboelastography, 201
Thrombogenetic potential, 188
Thrombus formation, 187
Thylakoid membrane lipid, 42
Time-of-flight secondary ion mass spectrum, 195
Topology, 39
Transbilayer movement, 110
Transbilayer movement, 110
Transfer protein, 12
Transient dichroism, 139, 140, 143
Transient membrane–protein complex, 154
Transition metal complex, 30, 31
Transition temperature, 21
Translational diffusion
    lipid matrix, 4
    phosphatidylcholine, 73
Transmembrane helix, 165
Triplet probe, 137–140

Unordered conformation, 156
Unsaturation
    fatty acid, 31
    lipid, 21, 30, 38
    phospholipid, 19–27, 34

Van der Waals interaction, 107
Vibrational spectroscopy, 93, 96

Water
    absorption, 168
    diffusion coefficient, 72
    –insoluble homogenous catalyst, 31–33
    –soluble homogenous catalyst, 33–36

–sugar hydroxyl group
  interaction, 122
uptake per mol lipid, 57
Wedge-like geometry, 73
Wide-angle X-ray diffraction, 83
Wilkinson's catalyst, 32, 44

X-ray diffraction
  membrane protein, 165, 174
  L_a, 58
X-ray photoelectron spectrum of
  DPPC surface, 194

Zone refining, 78